www.EffortlessMath.com

... So Much More Online!

✓ FREE Math lessons

✓ More Math learning books!

✓ Mathematics Worksheets

✓ Online Math Tutors

Need a PDF version of this book?

Please visit www.EffortlessMath.com

Comprehensive ISEE Middle Level Math Practice Book 2020 - 2021

Complete Coverage of all ISEE Middle Level Math Concepts + 2 Full-Length ISEE Middle Level Math Tests

By

Reza Nazari & Ava Ross

All inquiries should be addressed to:

info@effortlessMath.com

www.EffortlessMath.com

ISBN: 978-1-64612-351-3

Published by: Effortless Math Education

www.EffortlessMath.com

Visit www.EffortlessMath.com
for Online Math Practice

Description

Comprehensive ISEE Middle Level Math Practice Book 2020 - 2021, which reflects the 2020 - 2021 test guidelines, is a precious learning resource for ISEE Middle Level test-takers who need extra practice in math to raise their ISEE Middle Level Math scores. Upon completion of this exercise book, you will have a solid foundation and sufficient practice to ace the ISEE Middle Level Math test. **This comprehensive practice book is your ticket to scoring higher on ISEE Middle Level Math.**

The updated version of this unique practice workbook represents extensive exercises, math problems, sample ISEE Middle Level questions, and quizzes with answers and detailed solutions to help you hone your math skills, overcome your exam anxiety, boost your confidence—and do your best to defeat the ISEE Middle Level exam on test day.

Comprehensive ISEE Middle Level Math Practice Book 2020 – 2021 includes many exciting and unique features to help you improve your test scores, including:

- ✓ Content 100% aligned with the 2020 ISEE Middle Level® test
- ✓ Complete coverage of all ISEE Middle Level Math concepts and topics which you will be tested
- ✓ Over 2,500 additional ISEE Middle Level math practice questions in both multiple-choice and grid-in formats with answers grouped by topic, so you can focus on your weak areas
- ✓ Abundant Math skill-building exercises to help test-takers approach different question types that might be unfamiliar to them
- ✓ 2 full-length practice tests (featuring new question types) with detailed answers

This ISEE Middle Level Math practice book and other Effortless Math Education books are used by thousands of students each year to help them review core content areas, brush-up in math, discover their strengths and weaknesses, and achieve their best scores on the ISEE Middle Level test.

Contents

Chapter 1:

Fractions and Mixed Numbers

Math Topics that you'll learn in this Chapter:

- ✓ Simplifying Fractions

- ✓ Adding and Subtracting Fractions

- ✓ Multiplying and Dividing Fractions

- ✓ Adding Mixed Numbers

- ✓ Subtracting Mixed Numbers

- ✓ Multiplying Mixed Numbers

- ✓ Dividing Mixed Numbers

Simplifying Fractions

✎ *Simplify each fraction.*

1) $\frac{10}{15} =$

2) $\frac{8}{20} =$

3) $\frac{12}{42} =$

4) $\frac{5}{20} =$

5) $\frac{6}{18} =$

6) $\frac{18}{27} =$

7) $\frac{15}{55} =$

8) $\frac{24}{54} =$

9) $\frac{63}{72} =$

10) $\frac{40}{64} =$

11) $\frac{23}{46} =$

12) $\frac{35}{63} =$

13) $\frac{32}{36} =$

14) $\frac{81}{99} =$

15) $\frac{16}{64} =$

16) $\frac{14}{35} =$

17) $\frac{19}{38} =$

18) $\frac{18}{54} =$

19) $\frac{56}{70} =$

20) $\frac{40}{45} =$

21) $\frac{9}{90} =$

22) $\frac{20}{25} =$

23) $\frac{32}{48} =$

24) $\frac{7}{49} =$

25) $\frac{18}{48} =$

26) $\frac{54}{108} =$

Adding and Subtracting Fractions

✎ *Calculate and write the answer in lowest term.*

1) $\frac{1}{5} + \frac{1}{7} =$

2) $\frac{3}{7} + \frac{4}{5} =$

3) $\frac{3}{8} - \frac{1}{9} =$

4) $\frac{4}{5} - \frac{5}{9} =$

5) $\frac{2}{9} + \frac{1}{3} =$

6) $\frac{3}{10} + \frac{2}{5} =$

7) $\frac{9}{10} - \frac{4}{5} =$

8) $\frac{7}{9} - \frac{3}{7} =$

9) $\frac{3}{4} + \frac{1}{3} =$

10) $\frac{3}{8} + \frac{2}{5} =$

11) $\frac{3}{4} - \frac{2}{5} =$

12) $\frac{7}{9} - \frac{2}{3} =$

13) $\frac{4}{9} + \frac{5}{6} =$

14) $\frac{2}{3} + \frac{1}{4} =$

15) $\frac{9}{10} - \frac{3}{5} =$

16) $\frac{7}{12} - \frac{1}{2} =$

17) $\frac{4}{5} + \frac{2}{3} =$

18) $\frac{5}{7} + \frac{1}{5} =$

19) $\frac{5}{9} - \frac{2}{5} =$

20) $\frac{3}{5} - \frac{2}{9} =$

21) $\frac{7}{9} + \frac{1}{7} =$

22) $\frac{5}{8} + \frac{2}{3} =$

23) $\frac{4}{7} + \frac{2}{3} =$

24) $\frac{6}{7} - \frac{4}{9} =$

25) $\frac{4}{5} - \frac{2}{15} =$

26) $\frac{2}{9} + \frac{4}{5} =$

Multiplying and Dividing Fractions

✍ *Solve and write the answer in lowest term.*

1) $\frac{1}{2} \times \frac{4}{5} =$

2) $\frac{1}{5} \times \frac{6}{7} =$

3) $\frac{1}{3} \div \frac{1}{7} =$

4) $\frac{1}{7} \div \frac{3}{8} =$

5) $\frac{2}{3} \times \frac{4}{7} =$

6) $\frac{5}{7} \times \frac{3}{4} =$

7) $\frac{2}{5} \div \frac{3}{7} =$

8) $\frac{3}{7} \div \frac{5}{8} =$

9) $\frac{3}{8} \times \frac{4}{7} =$

10) $\frac{2}{9} \times \frac{6}{11} =$

11) $\frac{1}{10} \div \frac{3}{8} =$

12) $\frac{3}{10} \div \frac{4}{5} =$

13) $\frac{6}{7} \times \frac{4}{9} =$

14) $\frac{3}{7} \times \frac{5}{6} =$

15) $\frac{7}{9} \div \frac{6}{11} =$

16) $\frac{1}{15} \div \frac{2}{3} =$

17) $\frac{1}{13} \times \frac{1}{2} =$

18) $\frac{1}{12} \times \frac{4}{7} =$

19) $\frac{1}{15} \div \frac{4}{9} =$

20) $\frac{1}{16} \div \frac{1}{2} =$

21) $\frac{4}{7} \times \frac{5}{8} =$

22) $\frac{1}{11} \times \frac{4}{5} =$

23) $\frac{1}{18} \div \frac{5}{6} =$

24) $\frac{1}{15} \div \frac{3}{8} =$

25) $\frac{1}{11} \times \frac{3}{4} =$

26) $\frac{1}{14} \times \frac{2}{3} =$

Adding Mixed Numbers

✍ *Solve and write the answer in lowest terms.*

1) $3\frac{1}{5} + 2\frac{2}{9} =$

2) $1\frac{1}{7} + 5\frac{2}{5} =$

3) $4\frac{4}{5} + 1\frac{2}{7} =$

4) $2\frac{4}{7} + 2\frac{3}{5} =$

5) $1\frac{5}{6} + 1\frac{2}{5} =$

6) $3\frac{5}{7} + 1\frac{2}{9} =$

7) $3\frac{5}{8} + 2\frac{1}{3} =$

8) $1\frac{6}{7} + 3\frac{2}{9} =$

9) $2\frac{5}{9} + 1\frac{1}{4} =$

10) $3\frac{7}{9} + 2\frac{5}{6} =$

11) $2\frac{1}{10} + 2\frac{2}{5} =$

12) $1\frac{3}{10} + 3\frac{4}{5} =$

13) $3\frac{1}{12} + 2\frac{1}{3} =$

14) $5\frac{1}{11} + 1\frac{1}{2} =$

15) $3\frac{1}{21} + 2\frac{2}{3} =$

16) $4\frac{1}{24} + 1\frac{5}{8} =$

17) $2\frac{1}{25} + 3\frac{3}{5} =$

18) $3\frac{1}{15} + 2\frac{2}{10} =$

19) $5\frac{6}{7} + 2\frac{1}{3} =$

20) $2\frac{1}{8} + 3\frac{3}{4} =$

21) $2\frac{5}{7} + 2\frac{2}{21} =$

22) $4\frac{1}{6} + 1\frac{4}{5} =$

23) $3\frac{5}{6} + 1\frac{2}{7} =$

24) $2\frac{7}{8} + 3\frac{1}{3} =$

25) $3\frac{1}{17} + 1\frac{1}{2} =$

26) $1\frac{1}{18} + 1\frac{4}{9} =$

Subtracting Mixed Numbers

✍ *Solve and write the answer in lowest terms.*

1) $3\frac{2}{5} - 1\frac{2}{9} =$

2) $5\frac{3}{5} - 1\frac{1}{7} =$

3) $4\frac{2}{5} - 2\frac{2}{7} =$

4) $8\frac{3}{4} - 2\frac{1}{8} =$

5) $9\frac{5}{7} - 7\frac{4}{21} =$

6) $11\frac{7}{12} - 9\frac{5}{6} =$

7) $9\frac{5}{9} - 8\frac{1}{8} =$

8) $13\frac{7}{9} - 11\frac{3}{7} =$

9) $8\frac{7}{12} - 7\frac{3}{8} =$

10) $11\frac{5}{9} - 9\frac{1}{4} =$

11) $6\frac{5}{6} - 2\frac{2}{9} =$

12) $5\frac{7}{8} - 4\frac{1}{3} =$

13) $9\frac{5}{8} - 8\frac{1}{2} =$

14) $4\frac{9}{16} - 2\frac{1}{4} =$

15) $3\frac{2}{3} - 1\frac{2}{15} =$

16) $5\frac{1}{2} - 4\frac{2}{17} =$

17) $5\frac{6}{7} - 2\frac{1}{3} =$

18) $3\frac{3}{7} - 2\frac{2}{21} =$

19) $7\frac{3}{10} - 5\frac{2}{15} =$

20) $4\frac{5}{6} - 2\frac{2}{9} =$

21) $6\frac{3}{7} - 2\frac{2}{9} =$

22) $7\frac{4}{5} - 6\frac{3}{7} =$

23) $10\frac{2}{3} - 9\frac{5}{8} =$

24) $9\frac{3}{4} - 7\frac{4}{9} =$

25) $15\frac{4}{5} - 13\frac{12}{25} =$

26) $13\frac{5}{12} - 7\frac{5}{24} =$

Multiplying Mixed Numbers

✍ *Solve and write the answer in lowest terms.*

1) $1\frac{1}{8} \times 1\frac{3}{4} =$

2) $3\frac{1}{5} \times 2\frac{2}{7} =$

3) $2\frac{1}{8} \times 1\frac{2}{9} =$

4) $2\frac{3}{8} \times 2\frac{2}{5} =$

5) $1\frac{1}{2} \times 5\frac{2}{3} =$

6) $3\frac{1}{2} \times 6\frac{2}{3} =$

7) $9\frac{1}{2} \times 2\frac{1}{6} =$

8) $2\frac{5}{8} \times 8\frac{3}{5} =$

9) $3\frac{4}{5} \times 4\frac{2}{3} =$

10) $5\frac{1}{3} \times 2\frac{2}{7} =$

11) $6\frac{1}{3} \times 3\frac{3}{4} =$

12) $7\frac{2}{3} \times 1\frac{8}{9} =$

13) $8\frac{1}{2} \times 2\frac{1}{6} =$

14) $4\frac{1}{5} \times 8\frac{2}{3} =$

15) $3\frac{1}{8} \times 5\frac{2}{3} =$

16) $2\frac{2}{7} \times 6\frac{2}{5} =$

17) $2\frac{3}{8} \times 7\frac{2}{3} =$

18) $1\frac{7}{8} \times 8\frac{2}{3} =$

19) $9\frac{1}{2} \times 3\frac{1}{5} =$

20) $2\frac{5}{8} \times 4\frac{1}{3} =$

21) $6\frac{1}{3} \times 3\frac{2}{5} =$

22) $5\frac{3}{4} \times 2\frac{2}{7} =$

23) $9\frac{1}{4} \times 2\frac{1}{3} =$

24) $3\frac{3}{7} \times 7\frac{2}{5} =$

25) $4\frac{1}{4} \times 3\frac{2}{5} =$

26) $7\frac{2}{3} \times 3\frac{2}{5} =$

Dividing Mixed Numbers

✎ *Solve and write the answer in lowest terms.*

1) $9\frac{1}{2} \div 2\frac{3}{5} =$

2) $2\frac{3}{8} \div 1\frac{2}{5} =$

3) $5\frac{3}{4} \div 2\frac{2}{7} =$

4) $8\frac{1}{3} \div 4\frac{1}{4} =$

5) $7\frac{2}{5} \div 3\frac{3}{4} =$

6) $2\frac{4}{5} \div 3\frac{2}{3} =$

7) $8\frac{3}{5} \div 4\frac{3}{4} =$

8) $6\frac{3}{4} \div 2\frac{2}{9} =$

9) $5\frac{2}{7} \div 2\frac{2}{9} =$

10) $2\frac{2}{5} \div 3\frac{3}{5} =$

11) $4\frac{3}{7} \div 1\frac{7}{8} =$

12) $2\frac{5}{7} \div 2\frac{4}{5} =$

13) $8\frac{3}{5} \div 6\frac{1}{5} =$

14) $2\frac{5}{8} \div 1\frac{8}{9} =$

15) $5\frac{6}{7} \div 2\frac{3}{4} =$

16) $1\frac{3}{5} \div 2\frac{3}{8} =$

17) $5\frac{3}{4} \div 3\frac{2}{5} =$

18) $2\frac{3}{4} \div 3\frac{1}{5} =$

19) $3\frac{2}{3} \div 1\frac{2}{5} =$

20) $4\frac{1}{4} \div 2\frac{2}{3} =$

21) $3\frac{5}{6} \div 2\frac{4}{5} =$

22) $2\frac{1}{8} \div 1\frac{3}{4} =$

23) $5\frac{1}{2} \div 2\frac{2}{5} =$

24) $3\frac{4}{7} \div 2\frac{2}{3} =$

25) $2\frac{4}{5} \div 3\frac{5}{6} =$

26) $2\frac{3}{7} \div 3\frac{2}{3} =$

Answers – Chapter 1

Simplifying Fractions

1) $\frac{2}{3}$

2) $\frac{2}{5}$

3) $\frac{2}{7}$

4) $\frac{1}{4}$

5) $\frac{1}{3}$

6) $\frac{2}{3}$

7) $\frac{3}{11}$

8) $\frac{4}{9}$

9) $\frac{7}{8}$

10) $\frac{5}{8}$

11) $\frac{1}{2}$

12) $\frac{5}{9}$

13) $\frac{8}{9}$

14) $\frac{9}{11}$

15) $\frac{1}{4}$

16) $\frac{2}{5}$

17) $\frac{1}{2}$

18) $\frac{1}{3}$

19) $\frac{4}{5}$

20) $\frac{8}{9}$

21) $\frac{1}{10}$

22) $\frac{4}{5}$

23) $\frac{2}{3}$

24) $\frac{1}{7}$

25) $\frac{3}{8}$

26) $\frac{1}{2}$

Adding and Subtracting Fractions

1) $\frac{12}{35}$

2) $\frac{43}{35}$

3) $\frac{19}{72}$

4) $\frac{11}{45}$

5) $\frac{5}{9}$

6) $\frac{7}{10}$

25) $\frac{2}{3}$

7) $\frac{1}{10}$

8) $\frac{22}{63}$

9) $\frac{13}{12}$

10) $\frac{31}{40}$

11) $\frac{7}{20}$

12) $\frac{1}{9}$

26) $\frac{46}{45}$

13) $\frac{23}{18}$

14) $\frac{11}{12}$

15) $\frac{3}{10}$

16) $\frac{1}{12}$

17) $\frac{22}{15}$

18) $\frac{32}{35}$

19) $\frac{7}{45}$

20) $\frac{17}{45}$

21) $\frac{58}{63}$

22) $\frac{31}{24}$

23) $\frac{26}{21}$

24) $\frac{26}{63}$

Multiplying and Dividing Fractions

1) $\frac{2}{5}$

2) $\frac{6}{35}$

3) $\frac{7}{3}$

4) $\frac{8}{21}$

5) $\frac{8}{21}$

6) $\frac{15}{28}$

7) $\frac{14}{15}$

8) $\frac{24}{35}$

9) $\frac{3}{14}$

10) $\frac{4}{33}$

11) $\frac{4}{15}$

12) $\frac{3}{8}$

13) $\frac{8}{21}$

14) $\frac{5}{14}$

15) $\frac{77}{54}$

16) $\frac{1}{10}$

17) $\frac{1}{26}$

18) $\frac{1}{21}$

19) $\frac{3}{20}$

20) $\frac{1}{8}$

21) $\frac{5}{14}$ 23) $\frac{1}{15}$ 25) $\frac{3}{44}$

22) $\frac{4}{55}$ 24) $\frac{8}{45}$ 26) $\frac{1}{21}$

Adding Mixed Numbers

1) $5\frac{19}{45}$ 8) $5\frac{5}{63}$ 15) $5\frac{5}{7}$ 22) $5\frac{29}{30}$

2) $6\frac{19}{35}$ 9) $3\frac{29}{36}$ 16) $5\frac{2}{3}$ 23) $5\frac{5}{42}$

3) $6\frac{3}{35}$ 10) $6\frac{11}{18}$ 17) $5\frac{16}{25}$ 24) $6\frac{5}{24}$

4) $5\frac{6}{35}$ 11) $4\frac{1}{2}$ 18) $5\frac{4}{15}$ 25) $4\frac{19}{34}$

5) $3\frac{7}{30}$ 12) $5\frac{1}{10}$ 19) $8\frac{4}{21}$ 26) $2\frac{1}{2}$

6) $4\frac{59}{63}$ 13) $5\frac{5}{12}$ 20) $5\frac{7}{8}$

7) $5\frac{23}{24}$ 14) $6\frac{13}{22}$ 21) $4\frac{17}{21}$

Subtracting Mixed Numbers

1) $2\frac{8}{45}$ 8) $2\frac{22}{63}$ 15) $2\frac{8}{15}$ 22) $1\frac{13}{35}$

2) $4\frac{16}{35}$ 9) $1\frac{5}{24}$ 16) $1\frac{13}{34}$ 23) $1\frac{1}{24}$

3) $2\frac{4}{35}$ 10) $2\frac{11}{36}$ 17) $3\frac{11}{21}$ 24) $2\frac{11}{36}$

4) $6\frac{5}{8}$ 11) $4\frac{11}{18}$ 18) $1\frac{1}{3}$ 25) $2\frac{8}{25}$

5) $2\frac{11}{21}$ 12) $1\frac{13}{24}$ 19) $2\frac{1}{6}$ 26) $6\frac{5}{24}$

6) $1\frac{3}{4}$ 13) $1\frac{1}{8}$ 20) $2\frac{11}{18}$

7) $1\frac{31}{72}$ 14) $2\frac{5}{16}$ 21) $4\frac{13}{63}$

Multiplying Mixed Numbers

1) $1\frac{31}{32}$ 6) $23\frac{1}{3}$ 11) $23\frac{3}{4}$ 16) $14\frac{22}{35}$

2) $7\frac{11}{35}$ 7) $20\frac{7}{12}$ 12) $14\frac{13}{27}$ 17) $18\frac{5}{24}$

3) $2\frac{43}{72}$ 8) $22\frac{23}{40}$ 13) $18\frac{5}{12}$ 18) $16\frac{1}{4}$

4) $5\frac{7}{10}$ 9) $17\frac{11}{15}$ 14) $36\frac{2}{5}$ 19) $30\frac{2}{5}$

5) $8\frac{1}{2}$ 10) $12\frac{4}{21}$ 15) $17\frac{17}{24}$ 20) $11\frac{3}{8}$

21) $21\frac{8}{15}$

22) $13\frac{1}{7}$

23) $21\frac{7}{12}$

24) $25\frac{13}{35}$

25) $14\frac{9}{20}$

26) $26\frac{1}{15}$

Dividing Mixed Numbers

1) $3\frac{17}{26}$

2) $1\frac{39}{56}$

3) $2\frac{33}{64}$

4) $1\frac{49}{51}$

5) $1\frac{73}{75}$

6) $\frac{42}{55}$

7) $1\frac{77}{95}$

8) $3\frac{3}{80}$

9) $2\frac{53}{140}$

10) $\frac{2}{3}$

11) $2\frac{88}{105}$

12) $\frac{95}{98}$

13) $1\frac{12}{31}$

14) $1\frac{53}{136}$

15) $2\frac{10}{77}$

16) $\frac{64}{95}$

17) $1\frac{47}{68}$

18) $\frac{55}{64}$

19) $2\frac{13}{21}$

20) $1\frac{19}{32}$

21) $1\frac{31}{84}$

22) $1\frac{3}{14}$

23) $2\frac{7}{24}$

24) $1\frac{19}{56}$

25) $\frac{84}{115}$

26) $\frac{51}{77}$

Chapter 2:

Decimals

Math Topics that you'll learn in this Chapter:

- ✓ Comparing Decimals
- ✓ Rounding Decimals
- ✓ Adding and Subtracting Decimals
- ✓ Multiplying and Dividing Decimals

Comparing Decimals

✎ *Compare. Use >, =, and <*

1) 0.88 ☐ 0.088

2) 0.56 ☐ 0.57

3) 0.99 ☐ 0.89

4) 1.55 ☐ 1.65

5) 1.58 ☐ 1.75

6) 2.91 ☐ 2.85

7) 14.56 ☐ 1.456

8) 17.85 ☐ 17.89

9) 21.52 ☐ 21.052

10) 11.12 ☐ 11.03

11) 9.650 ☐ 9.65

12) 8.578 ☐ 8.568

13) 3.15 ☐ 0.315

14) 16.61 ☐ 16.16

15) 18.581 ☐ 8.991

16) 25.05 ☐ 2.505

17) 4.55 ☐ 4.65

18) 0.158 ☐ 1.58

19) 0.881 ☐ 0.871

20) 0.505 ☐ 0.510

21) 0.772 ☐ 0.777

22) 0.5 ☐ 0.500

23) 16.89 ☐ 15.89

24) 12.25 ☐ 12.35

25) 5.82 ☐ 5.69

26) 1.320 ☐ 1.032

27) 0.082 ☐ 0.088

28) 0.99 ☐ 0.099

29) 2.560 ☐ 1.950

30) 0.770 ☐ 0.707

31) 15.54 ☐ 1.554

32) 0.323 ☐ 0.332

Rounding Decimals

✎ *Round each number to the underlined place value.*

1) 2.814 =

2) 3.5_6_2 =

3) 12.1_2_5 =

4) 1_5_.5 =

5) 1.9_8_1 =

6) 14._2_15 =

7) 17.5_4_8 =

8) 25.5_0_8 =

9) 3_1_.089 =

10) 69._3_45 =

11) 9.4_5_7 =

12) 1_2_.901 =

13) 2.6_5_8 =

14) 32._5_65 =

15) 6.0_5_8 =

16) 98.1_0_8 =

17) 27._7_05 =

18) 3_6_.75 =

19) 9.0_8_ =

20) 7._1_85 =

21) 22.5_4_7 =

22) 66._0_98 =

23) 8_7_.75 =

24) 18._5_41 =

25) 10.2_5_8 =

26) 13._4_56 =

27) 71.0_8_4 =

28) 2_9_.23 =

29) 45._5_5 =

30) 9_1_.08 =

31) 8_3_.433 =

32) 74._6_4 =

Adding and Subtracting Decimals

✍ *Solve.*

1) $15.63 + 19.64 =$

2) $16.38 + 17.59 =$

3) $75.31 - 59.69 =$

4) $49.38 - 29.89 =$

5) $24.32 + 26.45 =$

6) $36.25 + 18.37 =$

7) $47.85 - 35.12 =$

8) $85.65 - 67.48 =$

9) $25.49 + 34.18 =$

10) $19.99 + 48.66 =$

11) $46.32 - 27.77 =$

12) $54.62 - 48.12 =$

13) $24.42 + 16.54 =$

14) $52.13 + 12.32 =$

15) $82.36 - 78.65 =$

16) $64.12 - 49.15 =$

17) $36.41 + 24.52 =$

18) $85.96 - 74.63 =$

19) $52.62 - 42.54 =$

20) $21.20 + 24.58 =$

21) $32.15 + 17.17 =$

22) $96.32 - 85.54 =$

23) $89.78 - 69.85 =$

24) $29.28 + 39.79 =$

25) $11.11 + 19.99 =$

26) $28.82 + 20.88 =$

27) $63.14 - 28.91 =$

28) $56.61 - 49.72 =$

29) $26.13 + 31.13 =$

30) $30.19 + 20.87 =$

31) $66.24 - 59.10 =$

32) $89.31 - 72.17 =$

Multiplying and Dividing Decimals

✎ **Solve.**

1) $11.2 \times 0.4 =$

2) $13.5 \times 0.8 =$

3) $42.2 \div 2 =$

4) $54.6 \div 6 =$

5) $23.1 \times 0.3 =$

6) $1.2 \times 0.7 =$

7) $5.5 \div 0.5 =$

8) $64.8 \div 8 =$

9) $1.4 \times 0.5 =$

10) $4.5 \times 0.3 =$

11) $88.8 \div 4 =$

12) $10.5 \div 5 =$

13) $2.2 \times 0.3 =$

14) $0.2 \times 0.52 =$

15) $95.7 \div 100 =$

16) $36.6 \div 6 =$

17) $3.2 \times 2 =$

18) $4.1 \times 0.5 =$

19) $68.4 \div 2 =$

20) $27.9 \div 9 =$

21) $3.5 \times 4 =$

22) $4.8 \times 0.5 =$

23) $6.4 \div 4 =$

24) $72.8 \div 0.8 =$

25) $1.8 \times 3 =$

26) $6.5 \times 0.2 =$

27) $93.6 \div 3 =$

28) $45.15 \div 0.5 =$

29) $13.2 \times 0.4 =$

30) $11.2 \times 5 =$

31) $7.2 \div 0.8 =$

32) $96.4 \div 0.2 =$

Answers – Chapter 2

Comparing Decimals

1) 0.88 > 0.088
2) 0.56 < 0.57
3) 0.99 > 0.89
4) 1.55 < 1.65
5) 1.58 < 1.75
6) 2.91 > 2.85
7) 14.56 > 1.456
8) 17.85 < 17.89
9) 21.52 > 21.052
10) 11.12 > 11.03
11) 9.650 = 9.65

12) 8.578 > 8.568
13) 3.15 > 0.315
14) 16.61 > 16.16
15) 18.581 > 8.991
16) 25.05 > 2.505
17) 4.55 < 4.65
18) 0.158 < 1.58
19) 0.881 > 0.871
20) 0.505 < 0.510
21) 0.772 < 0.777
22) 0.5 = 0.500

23) 16.89 > 15.89
24) 12.25 < 12.35
25) 5.82 > 5.69
26) 1.320 > 1.032
27) 0.082 < 0.088
28) 0.99 > 0.099
29) 2.560 > 1.950
30) 0.770 > 0.707
31) 15.54 > 1.554
32) 0.323 < 0.332

Rounding Decimals

1) 2.814 = 3
2) 3.562 = 3.56
3) 12.125 = 12.13
4) 15.5 = 16
5) 1.981 = 1.98
6) 14.215 = 14.2
7) 17.548 = 17.55
8) 25.508 = 25.51
9) 31.089 = 31
10) 69.345 = 69.3
11) 9.457 = 9.46

12) 12.901 = 13
13) 2.658 = 2.66
14) 32.565 = 32.6
15) 6.058 = 6.06
16) 98.108 = 98.11
17) 27.705 = 27.7
18) 36.75 = 37
19) 9.08 = 9.1
20) 7.185 = 7.2
21) 22.547 = 22.55
22) 66.098 = 66.1

23) 87.75 = 88
24) 18.541 = 18.5
25) 10.258 = 10.26
26) 13.456 = 13.5
27) 71.084 = 71.08
28) 29.23 = 29
29) 45.55 = 45.6
30) 91.08 = 91
33) 83.433 = 83
34) 74.64 = 74.6

Adding and Subtracting Decimals

1) 35.27
2) 33.97
3) 15.62
4) 19.49
5) 50.77
6) 54.62
7) 12.73
8) 18.17
9) 59.67
10) 68.65

11) 18.55
12) 6.5
13) 40.96
14) 64.45
15) 3.71
16) 14.97
17) 60.93
18) 11.33
19) 10.08
20) 45.78

21) 49.32
22) 10.78
23) 19.93
24) 69.07
25) 31.1
26) 49.7
27) 34.23
28) 6.89
29) 57.26
30) 51.06

31) 7.14   32) 17.14

Multiplying and Dividing Decimals

1) 4.48
2) 10.8
3) 21.1
4) 9.1
5) 6.93
6) 0.84
7) 1.1
8) 8.1
9) 0.7
10) 1.35
11) 22.2

12) 2.1
13) 0.66
14) 0.104
15) 0.957
16) 6.1
17) 6.4
18) 2.05
19) 34.2
20) 3.1
21) 14
22) 2.4

23) 1.6
24) 91
25) 5.4
26) 1.3
27) 31.2
28) 90.3
29) 5.28
30) 56
31) 9
32) 482

Chapter 3:

Integers and Order of Operations

Math Topics that you'll learn in this Chapter:

- ✓ Adding and Subtracting Integers

- ✓ Multiplying and Dividing Integers

- ✓ Order of Operations

- ✓ Integers and Absolute Value

Adding and Subtracting Integers

✎ *Solve.*

1) $-(8) + 13 =$

2) $17 - (-12 - 8) =$

3) $(-15) + (-4) =$

4) $(-14) + (-8) + 9 =$

5) $-(23) + 19 =$

6) $(-7 + 5) - 9 =$

7) $28 + (-32) =$

8) $(-11) + (-9) + 5 =$

9) $25 - (8 - 7) =$

10) $-(29) + 17 =$

11) $(-38) + (-3) + 29 =$

12) $15 - (-7 + 9) =$

13) $24 - (8 - 2) =$

14) $(-7 + 4) - 9 =$

15) $(-17) + (-3) + 9 =$

16) $(-26) + (-7) + 8 =$

17) $(-9) + (-11) =$

18) $8 - (-23 - 13) =$

19) $(-16) + (-2) =$

20) $25 - (7 - 4) =$

21) $23 + (-12) =$

22) $(-18) + (-6) =$

23) $17 - (-21 - 7) =$

24) $-(28) - (-16) + 5 =$

25) $(-9 + 4) - 8 =$

26) $(-28) + (-6) + 17 =$

27) $-(21) - (-15) + 9 =$

28) $(-31) + (-6) =$

29) $(-17) + (-11) + 14 =$

30) $(-29) + (-10) + 13 =$

31) $-(24) - (-12) + 5 =$

32) $8 - (-19 - 10) =$

Multiplying and Dividing Integers

✎ *Solve.*

1) $(-9) \times (-8) =$

2) $6 \times (-6) =$

3) $49 \div (-7) =$

4) $(-64) \div 8 =$

5) $(4) \times (-6) =$

6) $(-9) \times (-11) =$

7) $(10) \div (-5) =$

8) $144 \div (-12) =$

9) $(10) \times (-2) =$

10) $(-8) \times (-2) \times 5 =$

11) $(8) \div (-2) =$

12) $45 \div (-15) =$

13) $(5) \times (-7) =$

14) $(-6) \times (-5) \times 4 =$

15) $(12) \div (-6) =$

16) $(14) \div (-7) =$

17) $196 \div (-14) =$

18) $(27 - 13) \times (-2) =$

19) $125 \div (-5) =$

20) $66 \div (-6) =$

21) $(-6) \times (-5) \times 3 =$

22) $(15 - 6) \times (-3) =$

23) $(32 - 24) \div (-4) =$

24) $72 \div (-6) =$

25) $(-14 + 8) \times (-7) =$

26) $(-3) \times (-9) \times 3 =$

27) $84 \div (-12) =$

28) $(-12) \times (-10) =$

29) $25 \times (-4) =$

30) $(-3) \times (-5) \times 5 =$

31) $(15) \div (-3) =$

32) $(-18) \div (3) =$

Order of Operation

✎ *Calculate.*

1) $18 + (32 \div 4) =$

2) $(3 \times 8) \div (-2) =$

3) $67 - (4 \times 8) =$

4) $(-11) \times (8 - 3) =$

5) $(18 - 7) \times (6) =$

6) $(6 \times 10) \div (12 + 3) =$

7) $(13 \times 2) - (24 \div 6) =$

8) $(-5) + (4 \times 3) + 8 =$

9) $(4 \times 2^3) + (16 - 9) =$

10) $(3^2 \times 7) \div (-2 + 1) =$

11) $[-2(48 \div 2^3)] - 6 =$

12) $(-4) + (7 \times 8) + 18 =$

13) $(3 \times 7) + (16 - 7) =$

14) $[3^3 \times (48 \div 2^3)] \div (-2) =$

15) $(14 \times 3) - (3^4 \div 9) =$

16) $(96 \div 12) \times (-3) =$

17) $(48 \div 2^2) \times (-2) =$

18) $(56 \div 7) \times (-5) =$

19) $(-2^2) + (7 \times 9) - 21 =$

20) $(2^4 - 9) \times (-6) =$

21) $[4^3 \times (50 \div 5^2)] \div (-16) =$

22) $(3^2 \times 4^2) \div (-4 + 2) =$

23) $6^2 - (-6 \times 4) + 3 =$

24) $4^2 - (5^2 \times 3) =$

25) $(-4) + (12^2 \div 3^2) - 7^2 =$

26) $(3^2 \times 5) + (-5^2 - 9) =$

27) $2[(3^2 \times 5) \times (-6)] =$

28) $(11^2 - 2^2) - (-7^2) =$

29) $(2^3 \times 3) - (49 \div 7) =$

30) $3[(3^2 \times 5) + (25 \div 5)] =$

31) $(6^2 \times 5) \div (-5) =$

32) $2^2[(6^3 \div 12) - (3^4 \div 27)] =$

Integers and Absolute Value

✍ *Calculate.*

1) $5 - |8 - 12| =$

2) $|15| - \dfrac{|-16|}{4} =$

3) $\dfrac{|9 \times -6|}{18} \times \dfrac{|-24|}{8} =$

4) $|13 \times 3| + \dfrac{|-72|}{9} =$

5) $4 - |11 - 18| - |3| =$

6) $|18| - \dfrac{|-12|}{4} =$

7) $\dfrac{|5 \times -8|}{10} \times \dfrac{|-2|}{11} =$

8) $|9 \times 3| + \dfrac{|-36|}{4} =$

9) $|-42 + 7| \times \dfrac{|-2 \times 5|}{10} =$

10) $6 - |17 - 11| - |5| =$

11) $|13| - \dfrac{|-5|}{6} =$

12) $\dfrac{|9 \times -4|}{12} \times \dfrac{|-4|}{9} =$

13) $|-75 + 50| \times \dfrac{|-4 \times 5|}{5} =$

14) $\dfrac{|-26|}{13} \times \dfrac{|-32|}{8} =$

15) $14 - |8 - 18| - |-12| =$

16) $|29| - \dfrac{|-2|}{5} =$

17) $\dfrac{|3 \times 8|}{2} \times \dfrac{|-33|}{3} =$

18) $|-45 + 15| \times \dfrac{|-12 \times 5|}{6} =$

19) $\dfrac{|-50|}{5} \times \dfrac{|-77|}{11} =$

20) $12 - |2 - 7| - |15| =$

21) $|18| - \dfrac{|-45|}{15} =$

22) $\dfrac{|7 \times 8|}{4} \times \dfrac{|-4|}{12} =$

23) $\dfrac{|30 \times 2|}{3} \times |-12| =$

24) $\dfrac{|-36|}{9} \times \dfrac{|-80|}{8} =$

25) $|-35 + 8| \times \dfrac{|-9 \times 5|}{15} =$

26) $|19| - \dfrac{|-18|}{2} =$

27) $14 - |11 - 23| + |2| =$

28) $|-39 + 7| \times \dfrac{|-4 \times 6|}{3} =$

Answers – Chapter 3

Adding and Subtracting Integers

1) 5
2) 37
3) −19
4) −13
5) −4
6) −11
7) −4
8) −15
9) 24
10) −12
11) −12

12) 13
13) 18
14) −12
15) −11
16) −25
17) −20
18) 44
19) −18
20) 22
21) 11
22) −24

23) 45
24) −7
25) −13
26) −17
27) 3
28) −37
29) −14
30) −26
31) −7
32) 37

Multiplying and Dividing Integers

1) 72
2) −36
3) −7
4) −8
5) −24
6) 99
7) −2
8) −12
9) −20
10) 80
11) −4

12) −3
13) −35
14) 150
15) −2
16) −2
17) −14
18) −28
19) 25
20) −11
21) 90
22) −27

23) −2
24) −12
25) 42
26) 81
27) −7
28) 120
29) −100
30) 75
31) −5
32) −6

Order of Operation

1) 26
2) −12
3) 35
4) −55
5) 66
6) 4
7) 22
8) 15
9) 39
10) −63

11) −18
12) 70
13) 30
14) −81
15) 33
16) −24
17) −24
18) −40
19) 38
20) −42

21) −8
22) −72
23) 63
24) −59
25) −37
26) 11
27) −540
28) 166
29) 17
30) 150

31) −36   32) 60

Integers and Absolute Value

1) 1   11) 4   21) 15
2) 11   12) 15   22) 56
3) 9   13) 100   23) 240
4) 47   14) 8   24) 40
5) −6   15) −8   25) 81
6) 15   16) 25   26) 10
7) 8   17) 132   27) 4
8) 36   18) 300   28) 256
9) 35   19) 70
10) −5   20) −8

Chapter 4:

Ratios and Proportions

Math Topics that you'll learn in this Chapter:

- ✓ Simplifying Ratios

- ✓ Proportional Ratios

- ✓ Similarity and Ratios

Simplifying Ratios

✎ *Simplify each ratio.*

1) $3:27 = $ ___:___

2) $2:8 = $ ___:___

3) $\frac{4}{28} = -$

4) $\frac{16}{40} = -$

5) $10:30 = $ ___:___

6) $5:30 = $ ___:___

7) $\frac{34}{38} = -$

8) $\frac{45}{63} = -$

9) $10:45 = $ ___:___

10) $20:30 = $ ___:___

11) $\frac{40}{64} = -$

12) $\frac{10}{110} = -$

13) $8:12 = $ ___:___

14) $16:20 = $ ___:___

15) $\frac{24}{48} = -$

16) $\frac{21}{77} = -$

17) $8:24 = $ ___:___

18) $9 \text{ to } 36 = $ ___:___

19) $\frac{64}{72} = -$

20) $\frac{45}{60} = -$

21) $12:15 = $ ___:___

22) $18:54 = $ ___:___

23) $\frac{36}{54} = -$

24) $\frac{48}{104} = -$

25) $15:75 = $ ___:___

26) $16:48 = $ ___:___

27) $\frac{15}{65} = -$

28) $\frac{44}{52} = -$

Proportional Ratios

✎ *Solve each proportion for x.*

1) $\frac{4}{7} = \frac{16}{x}$, $x =$ ____

2) $\frac{4}{9} = \frac{x}{18}$, $x =$ ____

3) $\frac{3}{5} = \frac{24}{x}$, $x =$ ____

4) $\frac{3}{10} = \frac{x}{50}$, $x =$ ____

5) $\frac{3}{11} = \frac{15}{x}$, $x =$ ____

6) $\frac{6}{15} = \frac{x}{45}$, $x =$ ____

7) $\frac{6}{19} = \frac{12}{x}$, $x =$ ____

8) $\frac{7}{16} = \frac{x}{32}$, $x =$ ____

9) $\frac{18}{21} = \frac{54}{x}$, $x =$ ____

10) $\frac{13}{15} = \frac{39}{x}$, $x =$ ____

11) $\frac{9}{13} = \frac{72}{x}$, $x =$ ____

12) $\frac{8}{30} = \frac{x}{180}$, $x =$ ____

13) $\frac{3}{19} = \frac{9}{x}$, $x =$ ____

14) $\frac{1}{3} = \frac{x}{90}$, $x =$ ____

15) $\frac{25}{45} = \frac{x}{9}$, $x =$ ____

16) $\frac{1}{6} = \frac{9}{x}$, $x =$ ____

17) $\frac{7}{9} = \frac{63}{x}$, $x =$ ___

18) $\frac{54}{72} = \frac{x}{8}$, $x =$ ____

19) $\frac{32}{40} = \frac{4}{x}$, $x =$ ____

20) $\frac{21}{42} = \frac{x}{6}$, $x =$ ____

21) $\frac{56}{72} = \frac{7}{x}$, $x =$ ____

22) $\frac{1}{14} = \frac{x}{42}$, $x =$ ____

23) $\frac{5}{7} = \frac{75}{x}$, $x =$ ____

24) $\frac{30}{48} = \frac{x}{8}$, $x =$ ____

25) $\frac{36}{88} = \frac{9}{x}$, $x =$ ____

26) $\frac{62}{68} = \frac{x}{34}$, $x =$ ____

27) $\frac{42}{60} = \frac{x}{10}$, $x =$ ____

28) $\frac{8}{9} = \frac{x}{108}$, $x =$ ____

29) $\frac{46}{69} = \frac{x}{3}$, $x =$ ____

30) $\frac{99}{121} = \frac{x}{11}$, $x =$ ____

31) $\frac{19}{21} = \frac{x}{63}$, $x =$ ____

32) $\frac{11}{12} = \frac{x}{48}$, $x =$ ____

Create Proportion

✍ *State if each pair of ratios form a proportion.*

1) $\frac{5}{8}$ *and* $\frac{25}{50}$

2) $\frac{2}{11}$ *and* $\frac{4}{22}$

3) $\frac{2}{5}$ *and* $\frac{8}{20}$

4) $\frac{3}{11}$ *and* $\frac{9}{33}$

5) $\frac{5}{10}$ *and* $\frac{15}{30}$

6) $\frac{4}{13}$ *and* $\frac{8}{24}$

7) $\frac{6}{9}$ *and* $\frac{24}{36}$

8) $\frac{7}{12}$ *and* $\frac{14}{20}$

9) $\frac{3}{8}$ *and* $\frac{27}{72}$

10) $\frac{12}{20}$ *and* $\frac{36}{60}$

11) $\frac{11}{12}$ *and* $\frac{55}{60}$

12) $\frac{12}{15}$ *and* $\frac{24}{25}$

13) $\frac{15}{19}$ *and* $\frac{20}{38}$

14) $\frac{10}{14}$ *and* $\frac{40}{56}$

15) $\frac{11}{13}$ *and* $\frac{44}{39}$

16) $\frac{15}{16}$ *and* $\frac{30}{32}$

17) $\frac{17}{19}$ *and* $\frac{34}{48}$

18) $\frac{5}{18}$ *and* $\frac{15}{54}$

19) $\frac{3}{14}$ *and* $\frac{18}{42}$

20) $\frac{7}{11}$ *and* $\frac{14}{32}$

21) $\frac{8}{11}$ *and* $\frac{32}{44}$

22) $\frac{9}{13}$ *and* $\frac{18}{26}$

✍ *Solve.*

23) The ratio of boys to girls in a class is 5:6. If there are 25 boys in the class, how many girls are in that class? _____

24) The ratio of red marbles to blue marbles in a bag is 4:7. If there are 77 marbles in the bag, how many of the marbles are red? _____

25) You can buy 8 cans of green beans at a supermarket for $3.20. How much does it cost to buy 48 cans of green beans? _____

ISEE Middle Level Math Workbook 2020 - 2021

Similarity and Ratios

✏️ *Each pair of figures is similar. Find the missing side.*

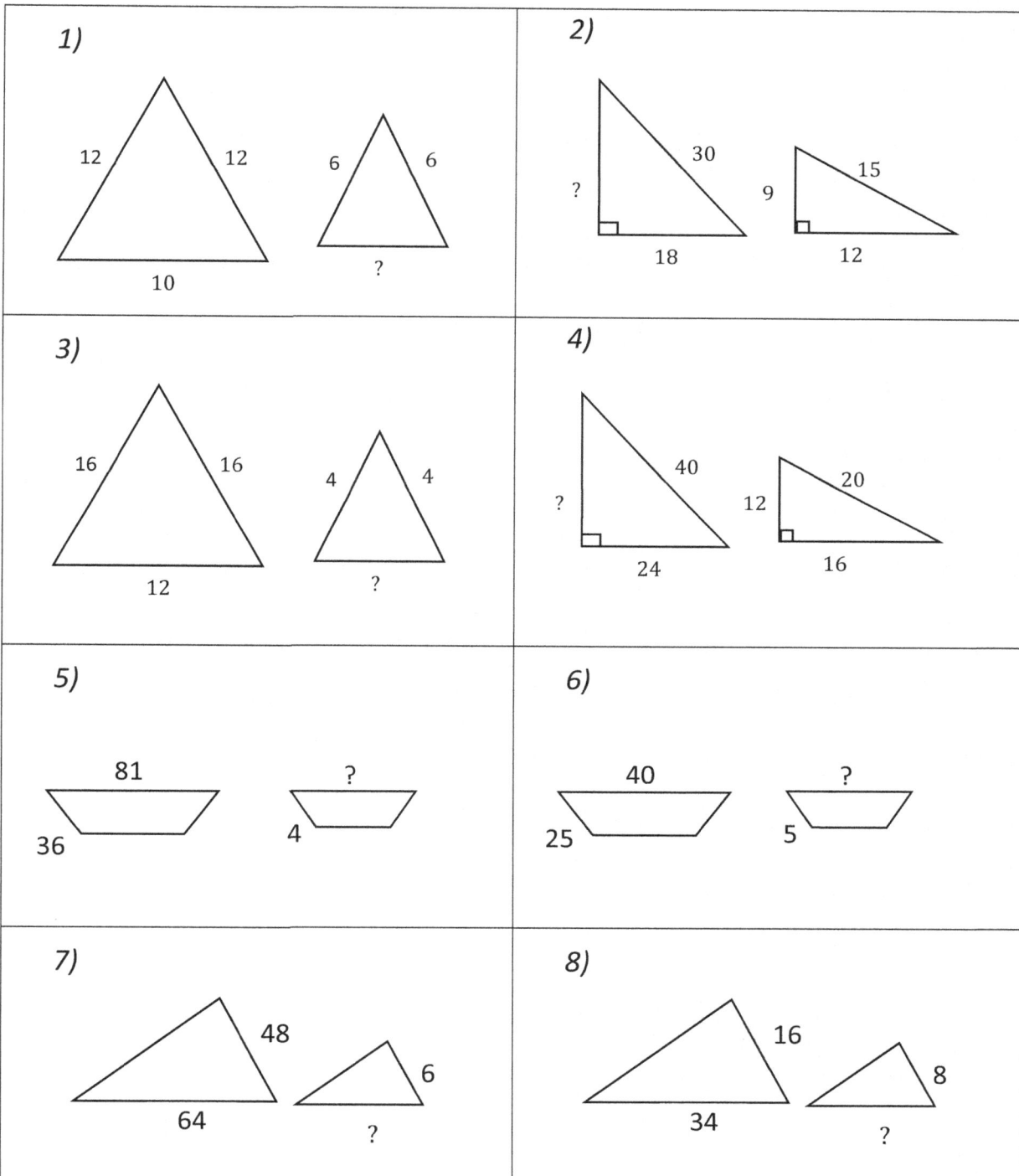

1) 12, 12, 10, 6, 6, ?

2) 30, ?, 18, 15, 9, 12

3) 16, 16, 12, 4, 4, ?

4) 40, ?, 24, 20, 12, 16

5) 81, 36, ?, 4

6) 40, 25, ?, 5

7) 48, 64, 6, ?

8) 16, 34, 8, ?

Answers – Chapter 4

Simplifying Ratios

1) $1:9$
2) $1:4$
3) $\frac{1}{7}$
4) $\frac{2}{5}$
5) $1:3$
6) $1:6$
7) $\frac{17}{19}$
8) $\frac{5}{7}$
9) $2:9$
10) $2:3$
11) $\frac{5}{8}$
12) $\frac{1}{11}$
13) $2:3$
14) $4:5$
15) $\frac{1}{2}$
16) $\frac{3}{11}$
17) $1:6$
18) 1 to 4
19) $\frac{8}{9}$
20) $\frac{3}{4}$
21) $4:5$
22) $1:3$
23) $\frac{2}{3}$
24) $\frac{6}{13}$
25) $1:5$
26) $1:3$
27) $\frac{3}{13}$
28) $\frac{11}{13}$

Proportional Ratios

1) $x = 28$
2) $x = 8$
3) $x = 40$
4) $x = 15$
5) $x = 55$
6) $x = 18$
7) $x = 38$
8) $x = 14$
9) $x = 63$
10) $x = 45$
11) $x = 104$
12) $x = 48$
13) $x = 57$
14) $x = 30$
15) $x = 5$
16) $x = 54$
17) $x = 81$
18) $x = 6$
19) $x = 5$
20) $x = 3$
21) $x = 9$
22) $x = 3$
23) $x = 105$
24) $x = 5$
25) $x = 22$
26) $x = 31$
27) $x = 7$
28) $x = 96$
29) $x = 2$
30) $x = 9$
31) $x = 57$
32) $x = 44$

Create Proportion

1) No
2) Yes
3) Yes
4) Yes
5) Yes
6) No
7) Yes
8) No
9) Yes
10) Yes
11) Yes
12) No
13) No
14) Yes
15) No
16) Yes
17) Yes
18) Yes
19) No
20) No
21) Yes

22) *Yes*

23) 30 *girls*
24) 28 *red marbles*
25) $19.20

Similarity and Ratios

1) 5
2) 24
3) 3
4) 32

5) 9
6) 8
7) 8
8) 17

Chapter 5:

Percentage

Math Topics that you'll learn in this Chapter:

✓ Percent Problems

✓ Percent of Increase and Decrease

✓ Simple Interest

✓ Discount, Tax and Tip

Percent Problems

✎ *Solve each problem.*

1) What is 5 percent of 300? ____

2) What is 15 percent of 600? ____

3) What is 12 percent of 450? ____

4) What is 30 percent of 240? ____

5) What is 60 percent of 850? ____

6) 63 is what percent of 300? ____%

7) 80 is what percent of 400? ____%

8) 70 is what percent of 700? ____%

9) 84 is what percent of 600? ___%

10) 90 is what percent of 300? ___%

11) 24 is what percent of 150? ___%

12) 12 is what percent of 80? ____%

13) 4 is what percent of 50? ____%

14) 110 is what percent of 500? _%

15) 16 is what percent of 400? __%

16) 39 is what percent of 300? ___%

17) 56 is what percent of 200? ___%

18) 30 is what percent of 500? ___%

19) 84 is what percent of 700? ___%

20) 40 is what percent of 500? __%

21) 26 is what percent of 100? __ %

22) 45 is what percent of 900? __%

23) 60 is what percent of 400? ____%

24) 18 is what percent of 900? ____%

25) 75 is what percent of 250? ____%

26) 27 is what percent of 900? ____%

27) 49 is what percent of 700? ____%

28) 81 is what percent of 900? ____%

29) 90 is what percent of 500? ____%

30) 82 is 20 percent of what number? ____

31) 14 is 35 percent of what number? ____

32) 90 is 6 percent of what number? ____

33) 80 is 40 percent of what number? ____

34) 90 is 15 percent of what number? ____

35) 28 is 7 percent of what number? ____

36) 54 is 18 percent of what number? ____

37) 72 is 24 percent of what number? ____

Percent of Increase and Decrease

✍ *Solve each percent of change word problem.*

1) Bob got a raise, and his hourly wage increased from $24 to $36. What is the percent increase? _____ %

2) The price of gasoline rose from $2.20 to $2.42 in one month. By what percent did the gas price rise? _____ %

3) In a class, the number of students has been increased from 30 to 39. What is the percent increase? _____ %

4) The price of a pair of shoes increases from $28 to $35. What is the percent increase? ____ %

5) In a class, the number of students has been decreased from 24 to 18. What is the percentage decrease? _____ %

6) Nick got a raise, and his hourly wage increased from $50 to $55. What is the percent increase? _____ %

7) A coat was originally priced at $80. It went on sale for $70.40. What was the percent that the coat was discounted? _____ %

8) The price of a pair of shoes increases from $8 to $12. What is the percent increase? ____ %

9) A house was purchased in 2002 for $180,000. It is now valued at $144,000. What is the rate (percent) of depreciation for the house? _____ %

10) The price of gasoline rose from $3.00 to $3.15 in one month. By what percent did the gas price rise? _____ %

Simple Interest

✍ *Determine the simple interest for these loans.*

1) $440 at 5% for 6 years. $___

2) $460 at 2.5% for 4 years. $_

3) $500 at 3% for 5 years. $___

4) $550 at 9% for 2 years. $___

5) $690 at 5% for 6 months. $___

6) $620 at 7% for 3 years. $___

7) $650 at 4.5% for 10 years. $___

8) $850 at 4% for 2 years. $___

9) $640 at 7% for 3 years. $___

10) $300 at 9% for 9 months. $___

11) $760 at 8% for 2 years. $__

12) $910 at 5% for 5 years. $___

13) $540 at 3% for 6 years. $___

14) $780 at 2.5% for 4 years. $___

15) $1,600 at 7% for 3 months. $___

16) $310 at 4% for 4 years. $___

17) $950 at 6% for 5 years. $___

18) $280 at 8% for 7 years. $___

19) $310 at 6% for 3 years. $___

20) $990 at 5% for 4 months. $___

21) $380 at 6% for 5 years. $___

22) $580 at 6% for 4 years. $___

23) $1,200 at 4% for5 years. $___

24) $3,100 at 5% for 6 years. $___

25) $5,200 at 8% for 2 years. $___

26) $1,400 at 4% for 3 years. $___

27) $300 at 3% for 8 months. $___

28) $150 at 3.5% for 4 years. $___

29) $170 at 6% for 2 years. $___

30) $940 at 8% for 5 years. $___

31) $960 at 1.5% for 8 years. $_

32) $240 at 5% for 4 months. $___

33) $280 at 2% for 5 years. $___

34) $880 at 3% for 2 years. $___

35) $2,200 at 4.5% for 2 years. $___

36) $2,400 at 7% for 3 years. $___

37) $1,800 at 5% for 6 months. $___

38) $190 at 4% for 2 years. $___

39) $560 at 7% for 4 years. $___

40) $720 at 8% for 2 years. $_

41) $780 at 5% for 8 years. $___

42) $880 at 6% for 3 months. $___

Discount, Tax and Tip

✎ *Find the missing values.*

1) Original price of a computer: $400

   Tax: 5%, Selling price: $_____

2) Original price of a sofa: $600

   Tax: 12%, Selling price: $_____

3) Original price of a table: $550

   Tax: 18%, Selling price: $_____

4) Original price of a cell phone: $700

   Tax: 20%, Selling price: $_____

5) Original price of a printer: $400

   Tax: 22%, Selling price: $_____

6) Original price of a computer: $600

   Tax: 15%, Selling price: $_____

7) Restaurant bill: $24.00

   Tip: 25%, Final amount: $_____

8) Original price of a cell phone: $300

   Tax: 8%, Selling price: $_____

9) Original price of a carpet: $800

   Tax: 25%, Selling price: $_____

10) Original price of a camera: $200

    Discount: 35%, Selling price: $_____

11) Original price of a dress: $500

    Discount: 10%, Selling price: $_____

12) Original price of a monitor: $400

    Discount: 5%, Selling price: $_____

13) Original price of a laptop: $900

    Discount: 20%, Selling price: $_____

14) Restaurant bill: $54.00

    Tip: 20%, Final amount: $_____

Answers – Chapter 5

Percent Problems

1) 15
2) 90
3) 54
4) 72
5) 510
6) 21%
7) 20%
8) 10%
9) 14%
10) 30%
11) 16%
12) 15%
13) 8%

14) 22%
15) 4%
16) 13%
17) 28%
18) 6%
19) 12%
20) 8%
21) 26%
22) 5%
23) 15%
24) 2%
25) 30%
26) 3%

27) 7%
28) 9%
29) 18%
30) 410
31) 40
32) 1,500
33) 200
34) 600
35) 400
36) 300
37) 300

Percent of Increase and Decrease - Answers

1) 50%
2) 10%
3) 30%
4) 25%

5) 25%
6) 10%
7) 12%
8) 50%

9) 20%
10) 5%

Simple Interest

2) $132
3) $46
4) $75
5) $99
6) $17.25
7) $130.20
8) $292.50
9) $68
10) $134.40
11) $20.25
12) $121.60
13) $227.50
14) $97.20
15) $78

16) $28
17) $49.60
18) $285
19) $156.80
20) $55.80
21) $198
22) $114
23) $139.20
24) $240
25) $930
26) $832
27) $168
28) $6
29) $21

30) $20.40
31) $376
32) $115.20
33) $4
34) $28
35) $52.80
36) $198
37) $504
38) $45
39) $15.20
40) $156.80
41) $115.20
42) $312
43) $13.20

Discount, Tax and Tip - Answers

1) $420
2) $672
3) $649
4) $840
5) $488
6) $690
7) $30.00

8) $324
9) $1,000
10) $130
11) $450
12) $380
13) $720
14) $64.80

Chapter 6:

Expressions and Variables

Math Topics that you'll learn in this Chapter:

✓ Simplifying Variable Expressions

✓ Simplifying Polynomial Expressions

✓ Evaluating One Variable

✓ Evaluating Two Variables

✓ The Distributive Property

Simplifying Variable Expressions

✎ *Simplify and write the answer.*

1) $3x + 5 + 2x =$

2) $7x + 3 - 3x =$

3) $-2 - x^2 - 6x^2 =$

4) $(-6)(8x - 4) =$

5) $3 + 10x^2 + 2x =$

6) $8x^2 + 6x + 7x^2 =$

7) $2x^2 - 5x - 7x =$

8) $x - 3 + 5 - 3x =$

9) $2 - 3x + 12 - 2x =$

10) $5x^2 - 12x^2 + 8x =$

11) $2x^2 + 6x + 3x^2 =$

12) $2x^2 - 2x - x =$

13) $2x^2 - (-8x + 6) = 2$

14) $4x + 6(2 - 5x) =$

15) $10x + 8(10x - 6) =$

16) $9(-2x - 6) - 5 =$

17) $32x - 4 + 23 + 2x =$

18) $8x - 12x - x^2 + 13 =$

19) $(-6)(8x - 4) + 10x =$

20) $14x - 5(5 - 8x) =$

21) $23x + 4(9x + 3) + 12 =$

22) $3(-7x + 5) + 20x =$

23) $12x - 3x(x + 9) =$

24) $7x + 5x(3 - 3x) =$

25) $5x(-8x + 12) + 14x =$

26) $40x + 12 + 2x^2 =$

27) $5x(x - 3) - 10 =$

28) $8x - 7 + 8x + 2x^2 =$

29) $7x - 3x^2 - 5x^2 - 3 =$

30) $4 + x^2 - 6x^2 - 12x =$

31) $12x + 8x^2 + 2x + 20 =$

32) $23 + 15x^2 + 8x - 4x^2 =$

Simplifying Polynomial Expressions

✎ *Simplify and write the answer.*

1) $(2x^3 + 5x^2) - (12x + 2x^2) =$ _____

2) $(-x^5 + 2x^3) - (3x^3 + 6x^2) =$ _____

3) $(12x^4 + 4x^2) - (2x^2 - 6x^4) =$ _____

4) $4x - 3x^2 - 2(6x^2 + 6x^3) =$ _____

5) $(2x^3 - 3) + 3(2x^2 - 3x^3) =$ _____

6) $4(4x^3 - 2x) - (3x^3 - 2x^4) =$ _____

7) $2(4x - 3x^3) - 3(3x^3 + 4x^2) =$ _____

8) $(2x^2 - 2x) - (2x^3 + 5x^2) =$ _____

9) $2x^3 - (4x^4 + 2x) + x^2 =$ _____

10) $x^4 - 9(x^2 + x) - 5x =$ _____

11) $(-2x^2 - x^4) + (4x^4 - x^2) =$ _____

12) $4x^2 - 5x^3 + 15x^4 - 12x^3 =$ _____

13) $2x^2 - 5x^4 + 14x^4 - 11x^3 =$ _____

14) $2x^2 + 5x^3 - 7x^2 + 12x =$ _____

15) $2x^4 - 5x^5 + 8x^4 - 8x^2 =$ _____

16) $5x^3 + 17x - 5x^2 - 2x^3 =$ _____

Evaluating One Variable

✍ *Evaluate each expression using the value given.*

1) $x = 3 \Rightarrow 6x - 9 =$

2) $x = 2 \Rightarrow 7x - 10 =$

3) $x = 1 \Rightarrow 5x + 2 =$

4) $x = 2 \Rightarrow 3x + 9 =$

5) $x = 4 \Rightarrow 4x - 8 =$

6) $x = 2 \Rightarrow 5x - 2x + 10 =$

7) $x = 3 \Rightarrow 2x - x - 6 =$

8) $x = 4 \Rightarrow 6x - 3x + 4 =$

9) $x = -2 \Rightarrow 4x - 6x - 5 =$

10) $x = -1 \Rightarrow 3x - 5x + 11 =$

11) $x = 1 \Rightarrow x - 7x + 12 =$

12) $x = 2 \Rightarrow 2(-3x + 4) =$

13) $x = 3 \Rightarrow 4(-5x - 2) =$

14) $x = 2 \Rightarrow 5(-2x - 4) =$

15) $x = -2 \Rightarrow 3(-4x - 5) =$

16) $x = 3 \Rightarrow 8x + 5 =$

17) $x = -3 \Rightarrow 12x + 9 =$

18) $x = -1 \Rightarrow 9x - 8 =$

19) $x = 2 \Rightarrow 16x - 10 =$

20) $x = 1 \Rightarrow 4x + 3 =$

21) $x = 5 \Rightarrow 7x - 2 =$

22) $x = 7 \Rightarrow 28 - x =$

23) $x = 3 \Rightarrow 5x - 10 =$

24) $x = 12 \Rightarrow 40 - 2x =$

25) $x = 2 \Rightarrow 11x - 2 =$

26) $x = 3 \Rightarrow 2x - x + 10 =$

Evaluating Two Variables

✎ *Evaluate each expression using the values given.*

1) $2x + 3y, x = 2, y = 3$

_____

2) $3x + 4y, x = -1, y = -2$

_____

3) $x + 6y, x = 3, y = 1$

_____

4) $2a - (15 - b), a = 2, b = 3$

_____

5) $4a - (6 - 3b), a = 1, b = 4$

_____

6) $a - (8 - 2b), a = 2, b = 5$

_____

7) $3z + 21 + 5k, z = 4, k = 1$

_____

8) $-7a + 4b, a = 6, b = 3$

_____

9) $-4a + 3b, a = 2, b = 4$

_____

10) $-6a + 6b, a = 4, b = 3$

_____

11) $-8a + 2b, a = 4, b = 6$

_____

12) $4x + 6y, x = 6, y = 3$

_____

13) $2x + 9y, x = 8, y = 1$

_____

14) $x - 7y, x = 9, y = 4$

_____

15) $5x - 4y, x = 6, y = 3$

_____

16) $2z + 14 + 8k, z = 4, k = 1$

_____

17) $6x + 3y, x = 3, y = 8$

_____

18) $5a - 6b, a = -3, b = -1$

_____

19) $8a + 4b, a = -4, b = 3$

_____

20) $-2a - b, a = 4, b = 9$

_____

21) $-7a + 3b, a = 4, b = 3$

_____

22) $-5a + 9b, a = 7, b = 1$

_____

The Distributive Property

✎ *Use the distributive property to simply each expression.*

1) $(-3)(12x + 3) =$

2) $(-4x + 5)(-6) =$

3) $13(-4x + 2) =$

4) $7(6 - 3x) =$

5) $(6 - 5x)(-4) =$

6) $9(8 - 2x) =$

7) $(-4x + 6)5 =$

8) $(-2x + 7)(-8) =$

9) $8(-4x + 7) =$

10) $(-9x + 5)(-3) =$

11) $8(-x + 9) =$

12) $7(2 - 6x) =$

13) $(-12x + 4)(-3) =$

14) $(-6)(-10x + 6) =$

15) $(-5)(5 - 11x) =$

16) $9(4 - 8x) =$

17) $(-6x + 2)7 =$

18) $(-9)(1 - 12x) =$

19) $(-3)(4 - 6x) =$

20) $(2 - 8x)(-2) =$

21) $20(2 - x) =$

22) $12(-4x + 3) =$

23) $15(2 - 3x) =$

24) $(-4x + 5)2 =$

25) $(-11x + 8)(-2) =$

26) $14(5 - 8x) =$

Answers – Chapter 6

Simplifying Variable Expressions

1) $5x + 5$
2) $4x + 3$
3) $-7x^2 - 2$
4) $-48x + 24$
5) $10x^2 + 2x + 3$
6) $15x^2 + 6x$
7) $2x^2 - 12x$
8) $-2x + 2$
9) $-5x + 14$
10) $-7x^2 + 8x$
11) $5x^2 + 6x$
12) $2x^2 - 3x$
13) $2x^2 + 8x - 6$
14) $-26x + 12$
15) $90x - 48$
16) $-18x - 59$

17) $34x + 19$
18) $-x^2 - 4x + 13$
19) $-38x + 24$
20) $54x - 25$
21) $59x + 24$
22) $-x + 15$
23) $-3x^2 - 15x$
24) $-15x^2 + 22x$
25) $-40x^2 + 74x$
26) $2x^2 + 40x + 12$
27) $5x^2 - 15x - 10$
28) $2x^2 + 16x - 7$
29) $-8x^2 + 7x - 3$
30) $-5x^2 - 12x + 4$
31) $8x^2 + 14x + 20$
32) $11x^2 + 8x + 23$

Simplifying Polynomial Expressions

1) $2x^3 + 3x^2 - 12x$
2) $-x^5 - x^3 - 6x^2$
3) $18x^4 + 2x^2$
4) $-12x^3 - 15x^2 + 4x$
5) $-7x^3 + 6x^2 - 3$
6) $2x^4 + 13x^3 - 8x$
7) $-15x^3 - 12x^2 + 8x$
14) $5x^3 - 5x^2 + 12x$
15) $-5x^5 + 10x^4 - 8x^2$

8) $-2x^3 - 3x^2 - 2x$
9) $-4x^4 + 2x^3 + x^2 - 2x$
10) $x^4 - 9x^2 - 14x$
11) $3x^4 - 3x^2$
12) $15x^4 - 17x^3 + 4x^2$
13) $9x^4 - 11x^3 + 2x^2$

16) $3x^3 - 5x^2 + 17x$

Evaluating One Variable

1) 9
2) 4
3) 7
4) 15
5) 8
6) 16
7) −3

8) 16
9) −1
10) 13
11) 6
12) −4
13) −68
14) −40

15) 9
16) 29
17) −27
18) −17
19) 22
20) 7
21) 33

22) 21
23) 5
24) 16
25) 20
26) 13

Evaluating Two Variables

1) 13
2) -11
3) 9
4) -8
5) 10
6) 4
7) 38
8) -30
9) 4
10) 6
11) -20
12) 42
13) 25
14) -19
15) 18
16) 30
17) 42
18) -9
19) -20
20) -17
21) -19
22) -26

The Distributive Property

1) $-36x - 9$
2) $24x - 30$
3) $-52x + 26$
4) $-21x + 42$
5) $20x - 24$
6) $-18x + 72$
7) $-20x + 30$
8) $16x - 56$
9) $-32x + 56$
10) $27x - 15$
11) $-8x + 72$
12) $-42x + 14$
13) $36x - 12$
14) $60x - 36$
15) $55x - 25$
16) $-72x + 36$
17) $-42x + 14$
18) $108x - 9$
19) $18x - 12$
20) $16x - 4$
21) $-20x + 40$
22) $-48x + 36$
23) $-45x + 30$
24) $-8x + 10$
25) $22x - 16$
26) $-112x + 70$

Chapter 7:

Equations and Inequalities

Math Topics that you'll learn in this Chapter:

- ✓ One–Step Equations

- ✓ Multi–Step Equations

- ✓ System of Equations

- ✓ Graphing Single–Variable Inequalities

- ✓ One–Step Inequalities

- ✓ Multi–Step Inequalities

One–Step Equations

✎ *Solve each equation for x.*

1) $x - 15 = 24 \Rightarrow x = $ _____

2) $18 = -6 + x \Rightarrow x = $ _____

3) $19 - x = 8 \Rightarrow x = $ _____

4) $x - 22 = 24 \Rightarrow x = $ _____

5) $24 - x = 17 \Rightarrow x = $ _____

6) $16 - x = 3 \Rightarrow x = $ _____

7) $x + 14 = 12 \Rightarrow x = $ _____

8) $26 + x = 8 \Rightarrow x = $ _____

9) $x + 9 = -18 \Rightarrow x = $ _____

10) $x + 21 = 11 \Rightarrow x = $ _____

11) $17 = -5 + x \Rightarrow x = $ _____

12) $x + 20 = 29 \Rightarrow x = $ _____

13) $x - 13 = 19 \Rightarrow x = $ _____

14) $x + 9 = -17 \Rightarrow x = $ _____

15) $x + 4 = -23 \Rightarrow x = $ _____

16) $16 = -9 + x \Rightarrow x = $ _____

17) $4x = 28 \Rightarrow x = $ _____

18) $21 = -7x \Rightarrow x = $ _____

19) $12x = -12 \Rightarrow x = $ _____

20) $13x = 39 \Rightarrow x = $ _____

21) $8x = -16 \Rightarrow x = $ _____

22) $\frac{x}{2} = -5 \Rightarrow x = $ _____

23) $\frac{x}{9} = 6 \Rightarrow x = $ _____

24) $27 = \frac{x}{5} \Rightarrow x = $ _____

25) $\frac{x}{4} = -3 \Rightarrow x = $ _____

26) $x \div 8 = 7 \Rightarrow x = $ _____

27) $x \div 2 = -3 \Rightarrow x = $ _____

28) $4x = 48 \Rightarrow x = $ _____

29) $9x = 72 \Rightarrow x = $ _____

30) $8x = -32 \Rightarrow x = $ _____

31) $80 = -10x \Rightarrow x = $ _____

Multi –Step Equations

✎ *Solve each equation.*

1) $3x - 8 = 13 \Rightarrow x = $ ____

2) $23 = -(x - 5) \Rightarrow x = $ ____

3) $-(8 - x) = 15 \Rightarrow x = $ ____

4) $29 = -x + 12 \Rightarrow x = $ ____

5) $2(3 - 2x) = 10 \Rightarrow x = $ ____

6) $3x - 3 = 15 \Rightarrow x = $ ____

7) $32 = -x + 15 \Rightarrow x = $ ____

8) $-(10 - x) = -13 \Rightarrow x = $ ____

9) $-4(7 + x) = 4 \Rightarrow x = $ ____

10) $23 = 2x - 8 \Rightarrow x = $ ____

11) $-6(3 + x) = 6 \Rightarrow x = $ ____

12) $-3 = 3x - 15 \Rightarrow x = $ ____

13) $-7(12 + x) = 7 \Rightarrow x = $ ____

14) $8(6 - 4x) = 16 \Rightarrow x = $ ____

15) $18 - 4x = -9 - x \Rightarrow x = $ ____

16) $6(4 - x) = 30 \Rightarrow x = $ ____

17) $15 - 3x = -5 - x \Rightarrow x = $ ____

18) $9(-7 - 3x) = 18 \Rightarrow x = $ ____

19) $16 - 2x = -4 - 7x \Rightarrow x = $ ____

20) $14 - 2x = 14 + x \Rightarrow x = $ ____

21) $21 - 3x = -7 - 10x \Rightarrow x = $ __

22) $8 - 2x = 11 + x \Rightarrow x = $ ____

23) $10 + 12x = -8 + 6x \Rightarrow x = $ ____

24) $25 + 20x = -5 + 5x \Rightarrow x = $ ____

25) $16 - x = -8 - 7x \Rightarrow x = $ ____

26) $17 - 3x = 13 + x \Rightarrow x = $ ____

27) $22 + 5x = -8 - x \Rightarrow x = $ ____

28) $-9(7 + x) = 9 \Rightarrow x = $ ____

29) $11 + 3x = -4 - 2x \Rightarrow x = $ ____

30) $13 - 2x = 3 - 3x \Rightarrow x = $ ____

31) $19 - x = -1 - 11x \Rightarrow x = $ ____

32) $12 - 2x = -2 - 4x \Rightarrow x = $ ____

System of Equations

Solve each system of equations.

1) $-x + y = 2$ $x =$
   $-2x + y = 3$ $y =$

2) $-5x + y = -3$ $x =$
   $3x - 8y = 24$ $y =$

3) $y = -5$ $x =$
   $4x - 5y = 13$

4) $3y = -6x + 8$ $x =$
   $5x - 4y = -3$ $y =$

5) $10x - 8y = -15$ $x =$
   $-6x + 4y = 13$ $y =$

6) $-3x - 4y = 5$ $x =$
   $x - 2y = 5$ $y =$

7) $5x - 12y = -19$ $x =$
   $-6x + 7y = 8$ $y =$

8) $5x - 7y = -2$ $x =$
   $-x - 2y = -3$ $y =$

9) $-x + 3y = 3$ $x =$
   $-7x + 8y = -5$ $y =$

10) $-4x + 3y = -18$ $x =$
    $4x - y = 14$ $y =$

11) $6x - 7y = -8$ $x =$
    $-x - 4y = -9$ $y =$

12) $-3x + 2y = -16$ $x =$
    $4x - y = 13$ $y =$

13) $-5x + y = -3$ $x =$
    $3x - 8y = 24$ $y =$

14) $3x - 2y = 2$ $x =$
    $x - y = 2$ $y =$

15) $4x + 7y = 2$ $x =$
    $6x + 7y = 10$ $y =$

16) $5x + 7y = 18$ $x =$
    $-3x + 7y = -22$ $y =$

Graphing Single–Variable Inequalities

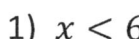 *Graph each inequality.*

1) $x < 6$

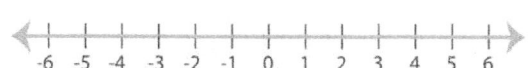

2) $x \geq 1$

3) $x \geq -6$

4) $x \leq -2$

5) $x > -1$

6) $3 > x$

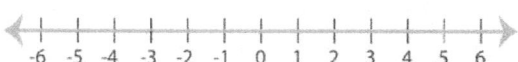

7) $2 \leq x$

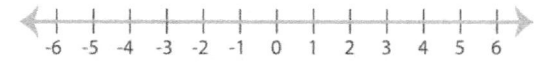

8) $x > 0$

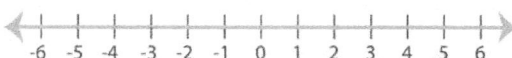

9) $-3 \leq x$

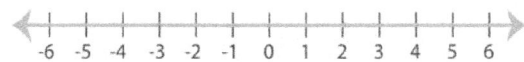

10) $-4 \leq x$

11) $x \leq 5$

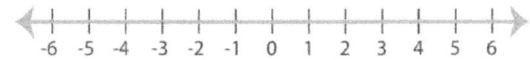

12) $0 \leq x$

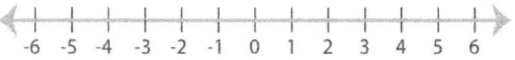

13) $-5 \leq x$

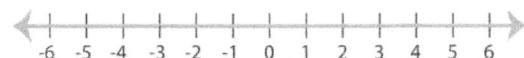

14) $x > -6$

One–Step Inequalities

✎ *Solve each inequality for x.*

1) $x - 10 < 22 \Rightarrow$ _____

2) $18 \leq -4 + x \Rightarrow$ _____

3) $x - 33 > 8 \Rightarrow$ _____

4) $x + 22 \geq 24 \Rightarrow$ _____

5) $x - 24 > 17 \Rightarrow$ _____

6) $x + 5 \geq 3 \Rightarrow x$_____

7) $x + 14 < 12 \Rightarrow$ _____

8) $26 + x \leq 8 \Rightarrow$ _____

9) $x + 9 \geq -18 \Rightarrow$ _____

10) $x + 24 < 11 \Rightarrow$ _____

11) $17 \leq -5 + x \Rightarrow$ _____

12) $x + 25 > 29 \Rightarrow x$_____

13) $x - 17 \geq 19 \Rightarrow$ _____

14) $x + 8 > -17 \Rightarrow$ _____

15) $x + 8 < -23 \Rightarrow$ _____

16) $16 \leq -5 + x \Rightarrow$ _____

17) $4x \leq 12 \Rightarrow$ _____

18) $28 \geq -7x \Rightarrow$ _____

19) $2x > -14 \Rightarrow$ _____

20) $13x \leq 39 \Rightarrow$ _____

21) $-8x > -16 \Rightarrow$ _____

22) $\frac{x}{2} < -6 \Rightarrow$ _____

23) $\frac{x}{6} > 6 \Rightarrow$ _____

24) $27 \leq \frac{x}{4} \Rightarrow$ _____

25) $\frac{x}{8} < -3 \Rightarrow$ _____

26) $6x \geq 18 \Rightarrow$ _____

27) $5x \geq -25 \Rightarrow$ _____

28) $4x > 48 \Rightarrow$ _____

29) $8x \leq 72 \Rightarrow$ _____

30) $-4x < -32 \Rightarrow$ _____

31) $40 > -10x \Rightarrow$ _____

Multi –Step Inequalities

✍ *Solve each inequality.*

1) $2x - 8 \leq 8 \rightarrow$ _____

2) $3 + 2x \geq 17 \rightarrow$ _____

3) $5 + 3x \geq 26 \rightarrow$ _____

4) $2x - 8 \leq 14 \rightarrow$ _____

5) $3x - 4 \leq 23 \rightarrow$ _____

6) $7x - 5 \leq 51 \rightarrow$ _____

7) $4x - 9 \leq 27 \rightarrow$ _____

8) $6x - 11 \leq 13 \rightarrow$ _____

9) $5x - 7 \leq 33 \rightarrow$ _____

10) $6 + 2x \geq 28 \rightarrow$ _____

11) $8 + 3x \geq 35 \rightarrow$ _____

12) $4 + 6x < 34 \rightarrow$ _____

13) $3 + 2x \geq 53 \rightarrow$ _____

14) $7 - 6x > 56 + x \rightarrow$ _____

15) $9 + 4x \geq 39 + 2x \rightarrow$ _____

16) $3 + 5x \geq 43 \rightarrow$ _____

17) $4 - 7x < 60 \rightarrow$ _____

18) $11 - 4x \geq 55 \rightarrow$ _____

19) $12 + x \geq 48 - 2x \rightarrow$ _____

20) $10 - 10x \leq -20 \rightarrow$ _____

21) $5 - 9x \geq -40 \rightarrow$ _____

22) $8 - 7x \geq 36 \rightarrow$ _____

23) $5 + 11x < 69 + 3x \rightarrow$ _____

24) $6 + 8x < 28 - 3x \rightarrow$ _____

25) $9 + 11x < 57 - x \rightarrow$ _____

26) $3 + 10x \geq 45 - 4x \rightarrow$ _____

Answers – Chapter 7

One–Step Equations

1) $x = 39$
2) $x = 24$
3) $x = 11$
4) $x = 46$
5) $x = 7$
6) $x = 13$
7) $x = 26$
8) $x = -18$
9) $x = -27$
10) $x = -10$
11) $x = 22$

12) $x = 9$
13) $x = 32$
14) $x = -26$
15) $x = -19$
16) $x = 25$
17) $x = 7$
18) $x = -3$
19) $x = -1$
20) $x = 3$
21) $x = -2$
22) $x = -10$

23) $x = 54$
24) $x = 135$
25) $x = -12$
26) $x = 56$
27) $x = -6$
28) $x = 12$
29) $x = 8$
30) $x = -4$
31) $x = -8$

Multi –Step Equations

1) $x = 7$
2) $x = -18$
3) $x = 23$
4) $x = -17$
5) $x = -1$
6) $x = 6$
7) $x = -17$
8) $x = -3$
9) $x = -8$
10) $x = 15$
11) $x = -4$

12) $x = 4$
13) $x = -13$
14) $x = 1$
15) $x = 9$
16) $x = -1$
17) $x = 10$
18) $x = -3$
19) $x = -4$
20) $x = 0$
21) $x = -4$
22) $x = -1$

23) $x = -3$
24) $x = -2$
25) $x = -4$
26) $x = 1$
27) $x = -5$
28) $x = -8$
29) $x = -3$
30) $x = -10$
31) $x = -2$
32) $x = -7$

System of Equations

1) $x = -1, y = 1$
2) $x = 0 , y = -3$
3) $x = -3$
4) $x = 1 , y = 2$
5) $x = -\frac{11}{2} , y = -5$
6) $x = 1 , y = -2$
7) $x = 1 , y = 2$
8) $x = 1 , y = 1$

9) $x = 3 , y = 2$
10) $x = 3 , y = -2$
11) $x = 1, y = 2$
12) $x = 2, y = -5$
13) $x = 0, y = -3$
14) $x = -2, y = -4$
15) $x = 4, y = -2$
16) $x = 5 , y = -1$

Graphing Single–Variable Inequalities

1) $x < 6$

2) $x \geq 1$

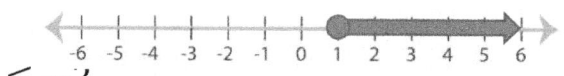

3) $x \geq -6$

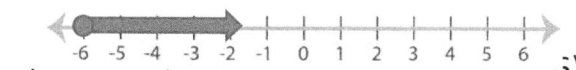

4) $x \leq -2$

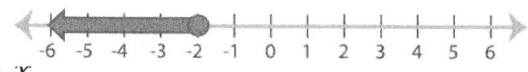

5) $x > -1$

6) $3 > x$

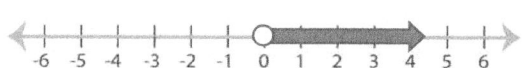

7) $2 \leq x$

8) $x > 0$

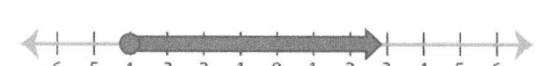

9) $-3 \leq x$

10) $-4 \leq x$

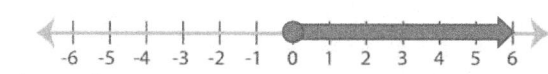

11) $x \leq 5$

12) $0 \leq x$

13) $-5 \leq x$

14) $x > -6$

One–Step Inequalities

1) $x < 32$
2) $22 \leq x$
3) $41 \leq x$
4) $x \geq 2$
5) $x > 41$
6) $x \geq -2$
7) $x < -2$
8) $x \leq -18$
9) $x \geq -27$
10) $x < -13$

11) $22 \leq x$
12) $x > 4$
13) $x \geq 36$
14) $x > -25$
15) $x < -31$
16) $21 \leq x$
17) $x \leq 3$
18) $-4 \leq x$
19) $x > -7$
20) $x \leq 3$
21) $x < 2$

22) $x < -12$
23) $x > 36$
24) $108 \leq x$
25) $x < -24$
26) $x \geq 3$
27) $x \geq -5$
28) $x > 12$
29) $x \leq 9$
30) $x > 8$
31) $-4 < x$

Multi –Step Inequalities

1) $x \leq 8$
2) $x \geq 7$
3) $x \geq 7$
4) $x \leq 11$
5) $x \leq 9$
6) $x \leq 8$
7) $x \leq 9$
8) $x \leq 4$
9) $x \leq 8$

10) $x \geq 11$
11) $x \geq 9$
12) $x < 5$
13) $x \geq 25$
14) $x < -7$
15) $x \geq 15$
16) $x \geq 8$
17) $x > -8$
18) $x \leq -11$

19) $x \geq 12$
20) $x \geq 3$
21) $x \leq 5$
22) $x \leq -4$
23) $x < 8$
24) $x < 2$
25) $x < 4$
26) $x \geq 3$

Chapter 8:

Exponents and Variables

Math Topics that you'll learn in this Chapter:

- ✓ Multiplication Property of Exponents

- ✓ Division Property of Exponents

- ✓ Powers of Products and Quotients

- ✓ Zero and Negative Exponents

- ✓ Negative Exponents and Negative Bases

- ✓ Scientific Notation

- ✓ Radicals

Multiplication Property of Exponents

✍ *Simplify and write the answer in exponential form.*

1) $2 \times 2^2 =$

2) $5^3 \times 5 =$

3) $3^2 \times 3^2 =$

4) $4^2 \times 4^2 =$

5) $7^3 \times 7^2 \times 7 =$

6) $2 \times 2^2 \times 2^2 =$

7) $5^3 \times 5^2 \times 5 \times 5 =$

8) $2x \times x =$

9) $x^3 \times x^2 =$

10) $x^4 \times x^4 =$

11) $x^2 \times x^2 \times x^2 =$

12) $6x \times 6x =$

13) $2x^2 \times 2x^2 =$

14) $3x^2 \times x =$

15) $4x^4 \times 4x^4 \times 4x^4 =$

16) $2x^2 \times x^2 =$

17) $x^4 \times 3x =$

18) $x \times 2x^2 =$

19) $5x^4 \times 5x^4 =$

20) $2yx^2 \times 2x =$

21) $3x^4 \times y^2x^4 =$

22) $y^2x^3 \times y^5x^2 =$

23) $4yx^3 \times 2x^2y^3 =$

24) $6x^2 \times 6x^3y^4 =$

25) $3x^4y^5 \times 7x^2y^3 =$

26) $7x^2y^5 \times 9xy^3 =$

27) $7xy^4 \times 4x^3y^3 =$

28) $3x^5y^3 \times 8x^2y^3 =$

29) $3x \times y^5x^3 \times y^4 =$

30) $yx^2 \times 2y^2x^2 \times 2xy =$

31) $4yx^4 \times 5y^5x \times xy^3 =$

32) $7x^2 \times 10x^3y^3 \times 8yx^4 =$

Division Property of Exponents

✎ *Simplify and write the answer.*

1) $\dfrac{2^2}{2^3} =$

2) $\dfrac{2^4}{2^2} =$

3) $\dfrac{5^5}{5} =$

4) $\dfrac{3}{3^5} =$

5) $\dfrac{x}{x^3} =$

6) $\dfrac{3 \times 3^3}{3^2 \times 3^4} =$

7) $\dfrac{5^8}{5^3} =$

8) $\dfrac{5 \times 5^6}{5^2 \times 5^7} =$

9) $\dfrac{3^4 \times 3^7}{3^2 \times 3^8} =$

10) $\dfrac{5x}{10x^3} =$

11) $\dfrac{5x^3}{2x^5} =$

12) $\dfrac{18x^3}{14x^6} =$

13) $\dfrac{12x^3}{8xy^8} =$

14) $\dfrac{24xy^3}{4x^4y^2} =$

15) $\dfrac{21x^3y^9}{7xy^5} =$

16) $\dfrac{36\ ^2y^9}{4x^3} =$

17) $\dfrac{12x^4y^4}{10x^6y^7} =$

18) $\dfrac{12y^2x^{12}}{20yx^8} =$

19) $\dfrac{16x^4y}{9x^8y^2} =$

20) $\dfrac{5x^8y^2}{20x^5y^5} =$

Powers of Products and Quotients

✎ *Simplify and write the answer.*

1) $(4^2)^2 =$

2) $(6^2)^3 =$

3) $(2 \times 2^3)^4 =$

4) $(4 \times 4^4)^2 =$

5) $(3^3 \times 3^2)^3 =$

6) $(5^4 \times 5^5)^2 =$

7) $(2 \times 2^4)^2 =$

8) $(2x^6)^2 =$

9) $(11x^5)^2 =$

10) $(4x^2y^4)^4 =$

11) $(2x^4y^4)^3 =$

12) $(3x^2y^2)^2 =$

13) $(3x^4y^3)^4 =$

14) $(2x^6y^8)^2 =$

15) $(12x^3x)^3 =$

16) $(5x^9x^6)^3 =$

17) $(5x^{10}y^3)^3 =$

18) $(14x^3x^3)^2 =$

19) $(3x^3 . 5x)^2 =$

20) $(10x^{11}y^3)^2 =$

21) $(9x^7y^5)^2 =$

22) $(4x^4y^6)^5 =$

23) $(3x . 4y^3)^2 =$

24) $\left(\frac{6x}{x^2}\right)^2 =$

25) $\left(\frac{x^5y^5}{x^2y^2}\right)^3 =$

26) $\left(\frac{24}{4x^6}\right)^2 =$

27) $\left(\frac{x^5}{x^7y^2}\right)^2 =$

28) $\left(\frac{xy^2}{x^2y^3}\right)^3 =$

29) $\left(\frac{4xy^4}{x^5}\right)^2 =$

30) $\left(\frac{xy^4}{5xy^2}\right)^3 =$

Zero and Negative Exponents

✎ *Evaluate the following expressions.*

1) $1^{-1} =$

2) $2^{-2} =$

3) $0^{15} =$

4) $1^{-10} =$

5) $8^{-1} =$

6) $8^{-2} =$

7) $2^{-4} =$

8) $10^{-2} =$

9) $9^{-2} =$

10) $3^{-3} =$

11) $7^{-3} =$

12) $3^{-4} =$

13) $6^{-3} =$

14) $5^{-3} =$

15) $22^{-1} =$

16) $4^{-4} =$

17) $5^{-4} =$

18) $15^{-2} =$

19) $4^{-5} =$

20) $9^{-3} =$

21) $3^{-5} =$

22) $5^{-4} =$

23) $12^{-3} =$

24) $15^{-3} =$

25) $20^{-3} =$

26) $50^{-2} =$

27) $18^{-3} =$

28) $24^{-2} =$

29) $30^{-3} =$

30) $10^{-5} =$

31) $\left(\frac{1}{8}\right)^{-1}$

32) $\left(\frac{1}{5}\right)^{-2} =$

33) $\left(\frac{1}{7}\right)^{-2} =$

34) $\left(\frac{2}{3}\right)^{-2} =$

35) $\left(\frac{1}{5}\right)^{-3} =$

36) $\left(\frac{3}{4}\right)^{-2} =$

37) $\left(\frac{2}{5}\right)^{-2} =$

38) $\left(\frac{1}{2}\right)^{-8} =$

39) $\left(\frac{2}{5}\right)^{-3} =$

40) $\left(\frac{3}{7}\right)^{-2} =$

41) $\left(\frac{5}{6}\right)^{-3} =$

42) $\left(\frac{4}{9}\right)^{-2} =$

Negative Exponents and Negative Bases

✍ *Simplify and write the answer.*

1) $-3^{-1} =$

2) $-5^{-2} =$

3) $-2^{-4} =$

4) $-x^{-3} =$

5) $2x^{-1} =$

6) $-4x^{-3} =$

7) $-12x^{-5} =$

8) $-5x^{-2}y^{-3} =$

9) $20x^{-4}y^{-1} =$

10) $14a^{-6}b^{-7} =$

11) $-12x^2y^{-3} =$

12) $-\dfrac{25}{x^{-6}} =$

13) $-\dfrac{2x}{a^{-4}} =$

14) $\left(-\dfrac{1}{3x}\right)^{-2} =$

15) $\left(-\dfrac{3}{4x}\right)^{-2} =$

16) $-\dfrac{9}{a^{-7}b^{-2}} =$

17) $-\dfrac{5x}{x^{-3}} =$

18) $-\dfrac{a^{-3}}{b^{-2}} =$

19) $-\dfrac{8}{x^{-3}} =$

20) $\dfrac{5b}{-9c^{-4}} =$

21) $\dfrac{9ab}{a^{-3}b^{-1}} =$

22) $-\dfrac{15a^{-2}}{30b^{-3}} =$

23) $\dfrac{4ab^{-2}}{-3c^{-2}} =$

24) $\left(\dfrac{3a}{2c}\right)^{-2} =$

25) $\left(-\dfrac{5x}{3yz}\right)^{-3} =$

26) $\dfrac{11ab^{-2}}{-3c^{-2}} =$

27) $\left(-\dfrac{x^3}{x^4}\right)^{-2} =$

28) $\left(-\dfrac{x^{-2}}{3x^2}\right)^{-3} =$

Scientific Notation

✍ **Write each number in scientific notation.**

1) 0.113 =

2) 0.02 =

3) 7.5 =

4) 20 =

5) 60 =

6) 0.004 =

7) 78 =

8) 1,600 =

9) 1,450 =

10) 31,000 =

11) 2,000,000 =

12) 0.0000003 =

13) 554,000 =

14) 0.000725 =

15) 0.00034 =

16) 86,000,000 =

17) 62,000 =

18) 97,000,000 =

19) 0.0000045 =

20) 0.0019 =

✍ **Write each number in standard notation.**

21) 2×10^{-1} =

22) 8×10^{-2} =

23) 1.8×10^{3} =

24) 9×10^{-4} =

25) 1.7×10^{-2} =

26) 9×10^{3} =

27) 7×10^{5} =

28) 1.15×10^{4} =

29) 7×10^{-5} =

30) 8.3×10^{-5} =

Radicals

✎ *Simplify and write the answer.*

1) $\sqrt{0} =$ ____

2) $\sqrt{1} =$ ____

3) $\sqrt{4} =$ ____

4) $\sqrt{16} =$ ____

5) $\sqrt{9} =$ ____

6) $\sqrt{25} =$ ____

7) $\sqrt{49} =$ ____

8) $\sqrt{36} =$ ____

9) $\sqrt{64} =$ ____

10) $\sqrt{81} =$ ____

11) $\sqrt{121} =$ ____

12) $\sqrt{225} =$ ____

13) $\sqrt{144} =$ ____

14) $\sqrt{100} =$ ____

15) $\sqrt{256} =$ ____

16) $\sqrt{289} =$ ____

17) $\sqrt{324} =$ ____

18) $\sqrt{400} =$ ____

19) $\sqrt{900} =$ ____

20) $\sqrt{529} =$ ____

21) $\sqrt{361} =$ ____

22) $\sqrt{169} =$ ____

23) $\sqrt{196} =$ ____

24) $\sqrt{90} =$ ____

✎ *Evaluate.*

25) $\sqrt{6} \times \sqrt{6} =$

26) $\sqrt{5} \times \sqrt{5} =$

27) $\sqrt{8} \times \sqrt{8} =$

28) $\sqrt{2} + \sqrt{2} =$

29) $\sqrt{8} + \sqrt{8} =$

30) $6\sqrt{5} - 2\sqrt{5} =$

31) $\sqrt{25} \times \sqrt{16} =$

32) $\sqrt{25} \times \sqrt{64} =$

33) $\sqrt{81} \times \sqrt{25} =$

34) $5\sqrt{3} \times 2\sqrt{3} =$

35) $8\sqrt{2} \times 2\sqrt{2} =$

36) $6\sqrt{3} - \sqrt{12} =$

Answers – Chapter 8

Multiplication Property of Exponents

1) 2^3
2) 5^4
3) 3^4
4) 4^4
5) 7^6
6) 2^5
7) 5^7
8) $2x^2$
9) x^5
10) x^8
11) x^6

12) $36x^2$
13) $4x^4$
14) $3x^3$
15) $64x^{12}$
16) $2x^4$
17) $3x^5$
18) $2x^3$
19) $25x^8$
20) $4x^3y$
21) $3x^8y^2$
22) x^5y^7

23) $8x^5y^4$
24) $36x^5y^4$
25) $21x^6y^8$
26) $63x^3y^8$
27) $28x^4y^7$
28) $24x^7y^6$
29) $3x^4y^9$
30) $4x^5y^4$
31) $20x^6y^9$
32) $560x^9y^4$

Division Property of Exponents

1) $\frac{1}{2}$
2) 2^2
3) 5^4
4) $\frac{1}{3^4}$
5) $\frac{1}{x^2}$
6) $\frac{1}{3}$
7) 5^5
8) $\frac{1}{5^2}$

9) 3
10) $\frac{1}{2x^2}$
11) $\frac{5}{2x^2}$
12) $\frac{9}{7x^3}$
13) $\frac{3x^2}{2y^8}$
14) $\frac{6y}{x^3}$
15) $3x^2y^4$

16) $\frac{9y^9}{x}$
17) $\frac{6}{5x^2y^3}$
18) $\frac{3yx^4}{5}$
19) $\frac{16}{9x^4y}$
20) $\frac{x^3}{4y^3}$

Powers of Products and Quotients

1) 4^4
2) 6^6
3) 2^{16}
4) 4^{10}
5) 3^{15}
6) 5^{18}
7) 2^{10}
8) $4x^{12}$
9) $121x^{10}$

10) $256x^8y^{16}$
11) $8x^{12}y^{12}$
12) $9x^4y^4$
13) $81x^{16}y^{12}$
14) $4x^{12}y^{16}$
15) $1,728x^{12}$
16) $125x^{45}$
17) $125x^{30}y^9$
18) $196x^{12}$

19) $225x^8$
20) $100x^{22}y^6$
21) $81x^{14}y^{10}$
22) $1,024x^{20}y^{30}$
23) $144x^2y^6$
24) $\frac{36}{x^2}$
25) x^9y^9
26) $\frac{36}{x^{10}}$

27) $\dfrac{1}{x^4 y^4}$

28) $\dfrac{1}{x^3 y^3}$

29) $\dfrac{16y^8}{x^8}$

30) $\dfrac{y^6}{125}$

Zero and Negative Exponents

1) 1

2) $\dfrac{1}{4}$

3) 0

4) 1

5) $\dfrac{1}{8}$

6) $\dfrac{1}{64}$

7) $\dfrac{1}{16}$

8) $\dfrac{1}{100}$

9) $\dfrac{1}{81}$

10) $\dfrac{1}{27}$

11) $\dfrac{1}{343}$

12) $\dfrac{1}{81}$

13) $\dfrac{1}{216}$

14) $\dfrac{1}{125}$

15) $\dfrac{1}{22}$

16) $\dfrac{1}{256}$

17) $\dfrac{1}{625}$

18) $\dfrac{1}{225}$

19) $\dfrac{1}{1,024}$

20) $\dfrac{1}{729}$

21) $\dfrac{1}{243}$

22) $\dfrac{1}{625}$

23) $\dfrac{1}{144}$

24) $\dfrac{1}{3,375}$

25) $\dfrac{1}{8,000}$

26) $\dfrac{1}{2,500}$

27) $\dfrac{1}{5,832}$

28) $\dfrac{1}{576}$

29) $\dfrac{1}{27,000}$

30) $\dfrac{1}{100,000}$

31) 8

32) 25

33) 49

34) $\dfrac{9}{4}$

35) 125

36) $\dfrac{64}{27}$

37) $\dfrac{25}{4}$

38) 256

39) $\dfrac{125}{8}$

40) $\dfrac{49}{9}$

41) $\dfrac{216}{125}$

42) $\dfrac{81}{16}$

Negative Exponents and Negative Bases

1) $-\dfrac{1}{3}$

2) $-\dfrac{1}{25}$

3) $-\dfrac{1}{16}$

4) $-\dfrac{1}{x^3}$

5) $\dfrac{2}{x}$

6) $-\dfrac{4}{x^3}$

7) $-\dfrac{12}{x^5}$

8) $-\dfrac{5}{x^2 y^3}$

9) $\dfrac{20}{x^4 y}$

10) $\dfrac{14}{a^6 b^7}$

11) $-\dfrac{12x^2}{y^3}$

12) $-25x^6$

13) $-2xa^4$

14) $9x^2$

15) $\dfrac{16x^2}{9}$

16) $-9a^7 b^2$

17) $-5x^4$

18) $-\dfrac{b^2}{a^3}$

19) $-8x^3$

20) $-\dfrac{5bc^4}{9}$

21) $9a^4 b^2$

22) $-\dfrac{b^3}{2a^2}$

23) $-\dfrac{4ac^2}{3b^2}$

24) $\dfrac{4c^2}{9a^2}$

25) $-\dfrac{27y^3 z^3}{125x^3}$

26) $-\dfrac{11ac^2}{3b^2}$

27) x^2

28) $-27x^{12}$

Scientific Notation

1) 1.13×10^{-1}
2) 2×10^{-2}
3) 2.5×10^{0}
4) 2×10^{1}
5) 6×10^{1}
6) 4×10^{-3}
7) 7.8×10^{1}
8) 1.6×10^{3}
9) 1.45×10^{3}
10) 3.1×10^{4}
20) 1.9×10^{-3}
21) $= 0.2$
22) 0.08
23) $1,800$
24) 0.0009
25) 0.017

11) 2×10^{6}
12) 3×10^{-7}
13) 5.54×10^{5}
14) 7.25×10^{-4}
15) 3.4×10^{-4}
16) 8.6×10^{7}
17) 6.2×10^{4}
18) 9.7×10^{7}
19) 4.5×10^{-6}

26) $9,000$
27) $700,000$
28) $11,500$
29) 0.00007
30) 0.000083

Radicals

1) 0
2) 1
3) 2
4) 4
5) 3
6) 5
7) 7
8) 6

9) 8
10) 9
11) 11
12) 15
13) 12
14) 10
15) 16
16) 17

17) 18
18) 20
19) 30
20) 23
21) 19
22) 13
23) 14
24) $3\sqrt{10}$

25) 6
26) 5
27) 8
28) $2\sqrt{2}$

29) $2\sqrt{8} = 4\sqrt{2}$
30) $4\sqrt{5}$
31) 20
32) 40

33) 45
34) 30
35) 32
36) $4\sqrt{3}$

Chapter 9:

Geometry and Solid Figures

Math Topics that you'll learn in this Chapter:

- ✓ The Pythagorean Theorem
- ✓ Triangles
- ✓ Polygons
- ✓ Circles
- ✓ Trapezoids
- ✓ Cubes
- ✓ Rectangle Prisms
- ✓ Cylinder

The Pythagorean Theorem

✎ *Do the following lengths form a right triangle?*

1) _____
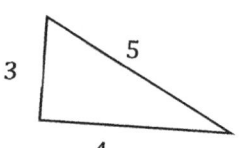

2) _____
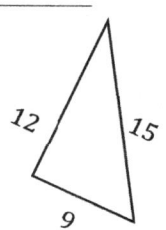

3) _____
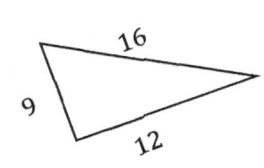

4) _____
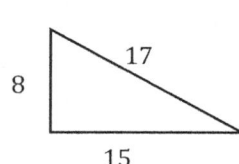

5) _____
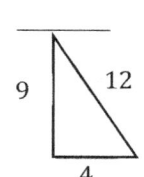

6) _____
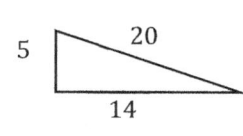

7) _____
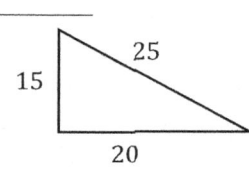

8) _____
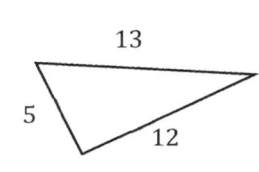

✎ *Find the missing side.*

9) _____
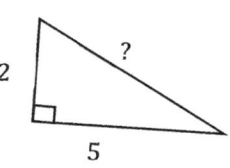

10) _____
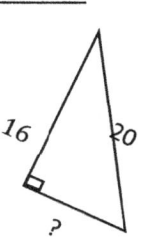

11) _____
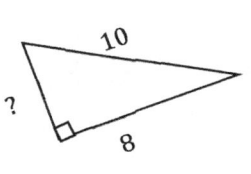

12) _____
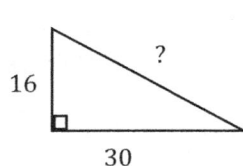

13) _____
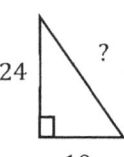

14) _____
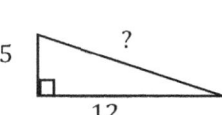

15) _____
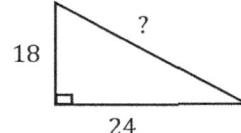

16) _____
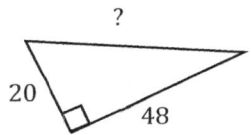

Triangles

✎ *Find the measure of the unknown angle in each triangle.*

1) _____

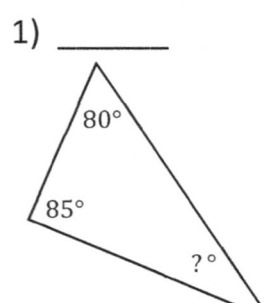

2) _____

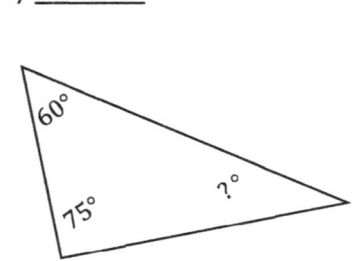

3) _____

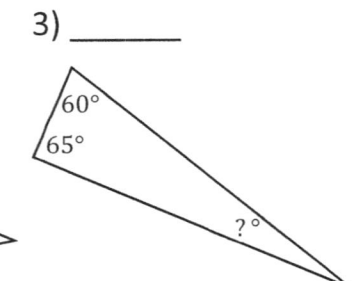

4) _____

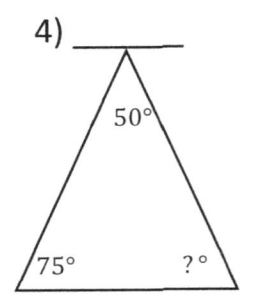

5) _____

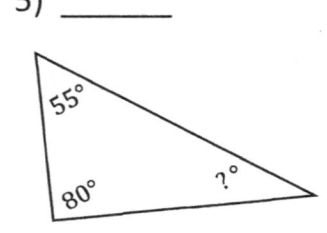

6) _____

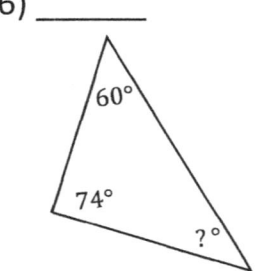

7) _____

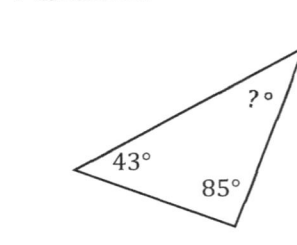

8) _____

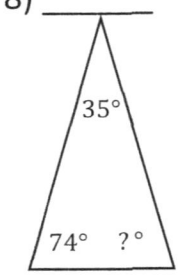

✎ *Find area of each triangle.*

9) _____

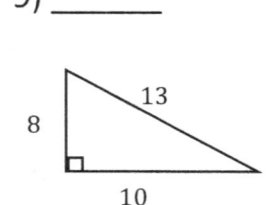

10) _____

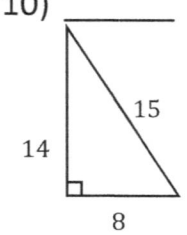

11) _____

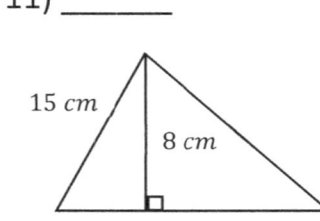

12) _____

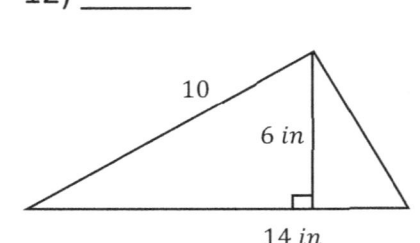

Polygons

✎ *Find the perimeter of each shape.*

1) (square) _____

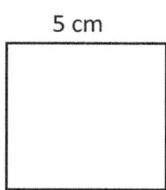

5 cm

2) _____

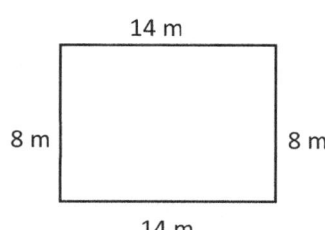

14 m
8 m   8 m
14 m

3) _____

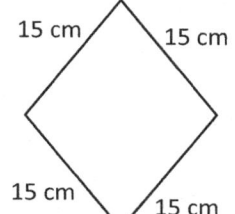

15 cm   15 cm
15 cm   15 cm

4) (square) _____

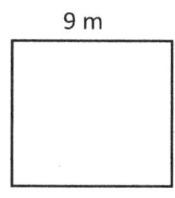

9 m

5) *(regular hexagon* _____

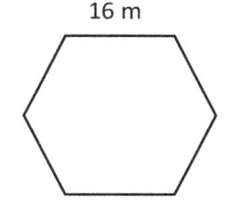

16 m

6) _____

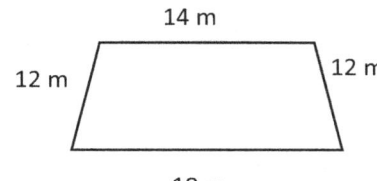

14 m
12 m   12 m
18 m

7) *(parallelogram* _____

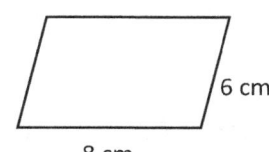

6 cm
8 cm

8) *(regular hexagon)* _____

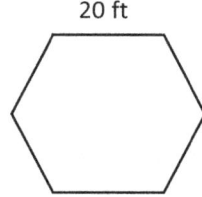

20 ft

9) _____

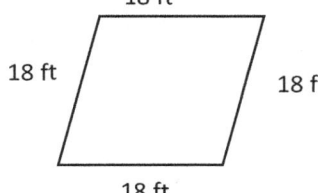

18 ft
18 ft   18 ft
18 ft

10) _____

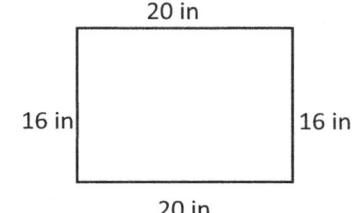

20 in
16 in   16 in
20 in

11) _____

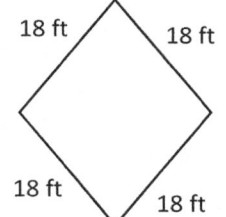

18 ft   18 ft
18 ft   18 ft

12) *(regular hexagon)* _____

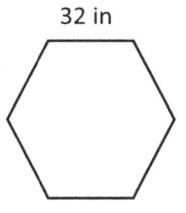

32 in

Circles

✎ **Find the Circumference of each circle.** ($\pi = 3.14$)

1) _____ 2) _____ 3) _____ 4) _____ 5) _____ 6) _____

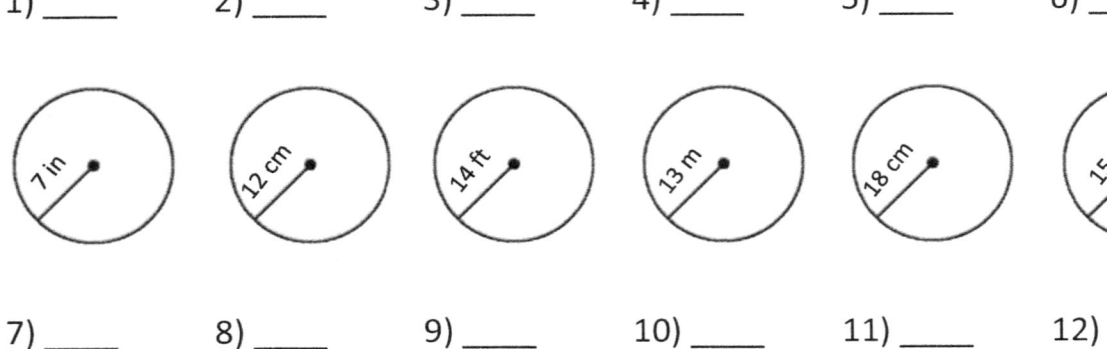

7) _____ 8) _____ 9) _____ 10) _____ 11) _____ 12) _____

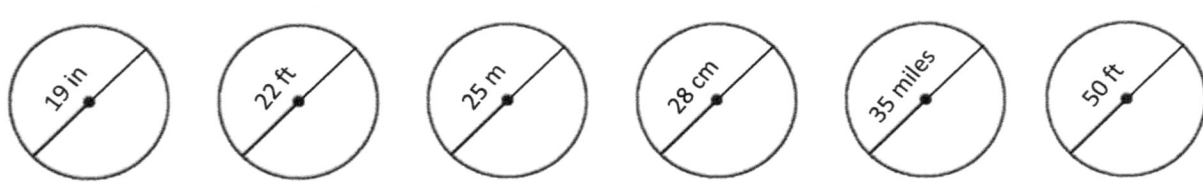

✎ **Complete the table below.** ($\pi = 3.14$)

	Radius	Diameter	Circumference	Area
Circle 1	2 inches	4 inches	12.56 inches	12.56 square inches
Circle 2		8 meters		
Circle 3				113.04 square ft
Circle 4			50.24 miles	
Circle 5		9 km		
Circle 6	7 cm			
Circle 7		10 feet		
Circle 8				615.44 square meters
Circle 9			81.64 inches	
Circle 10	12 feet			

ISEE Middle Level Math Workbook 2020 - 2021

Cubes

✎ *Find the volume of each cube.*

1) ___	2) ___	3) ___	4) ___	5) ___	6) ___
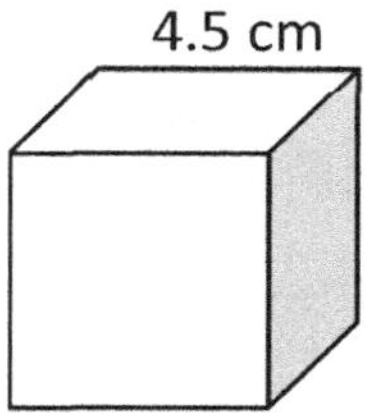 4.5 cm	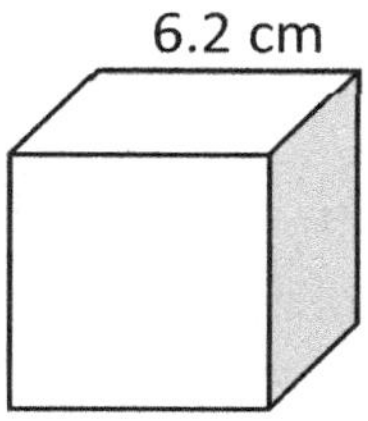 6.2 cm	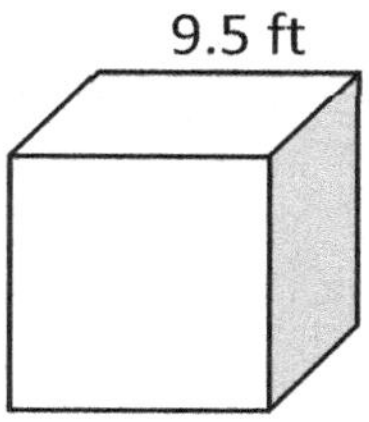 9.5 ft	11 m	13 in	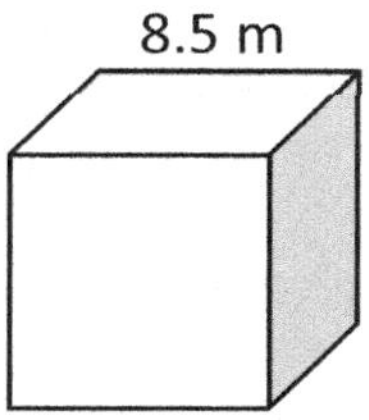 8.5 m

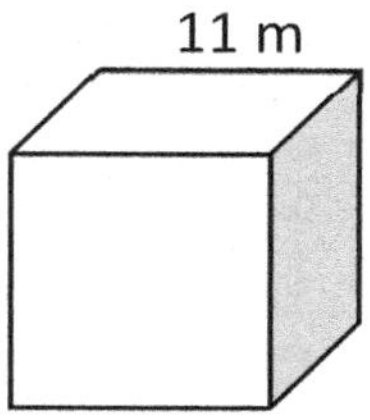

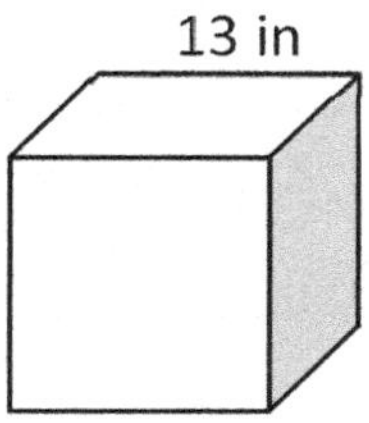

7) ___	8) ___	9) ___	10) ___	11) ___	12) ___
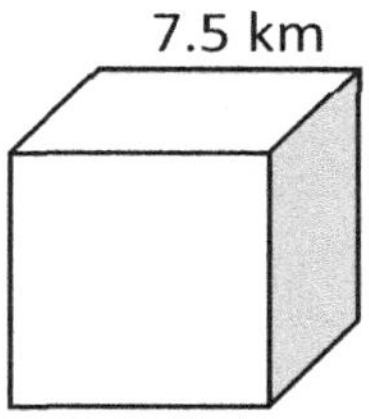 7.5 km	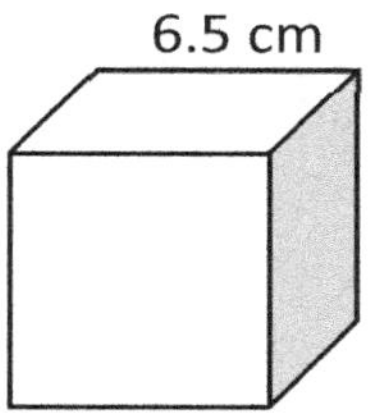 6.5 cm	16 ft	17 mm	30 in	40 km

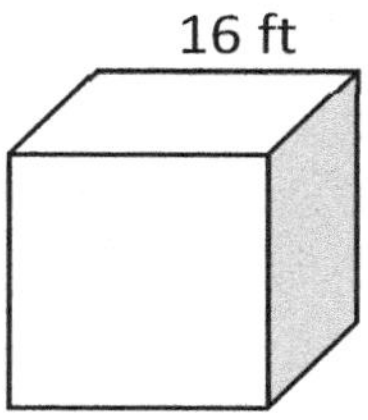

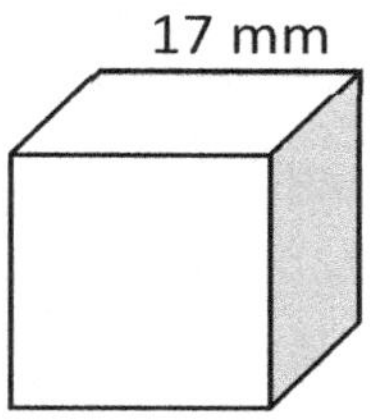

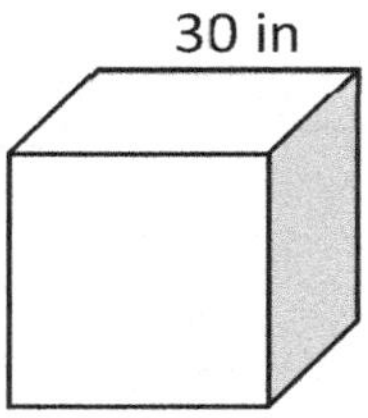

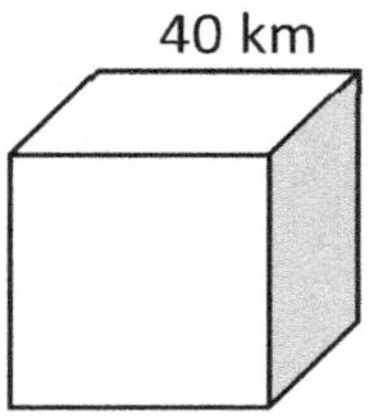

✎ *Find the surface area of each cube.*

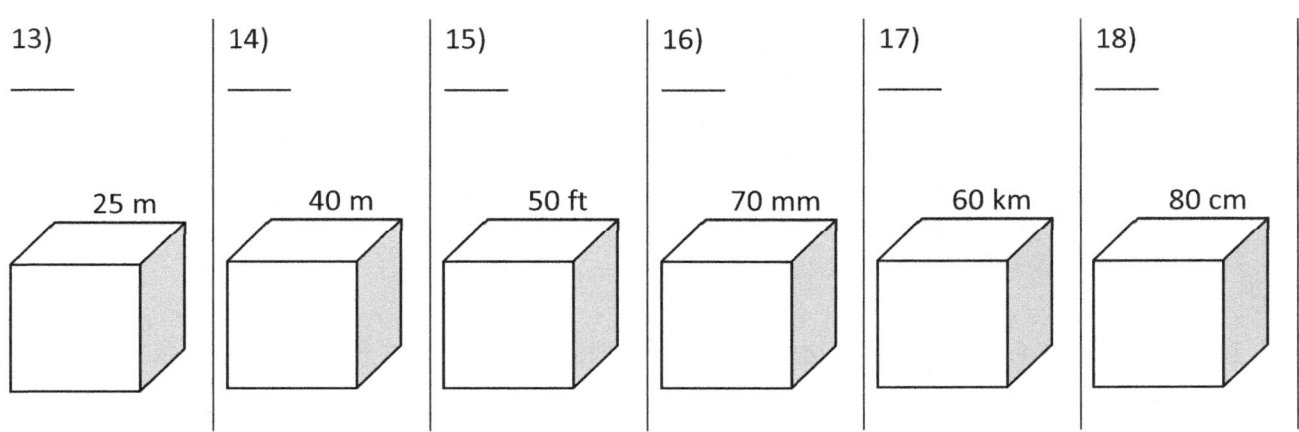

13) ___	14) ___	15) ___	16) ___	17) ___	18) ___
25 m	40 m	50 ft	70 mm	60 km	80 cm

Trapezoids

✏️ *Find the area of each trapezoid.*

1)  _____

2) _____

3) _____

4) _____

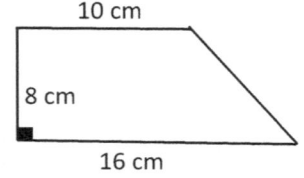

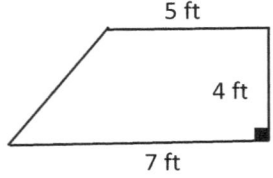

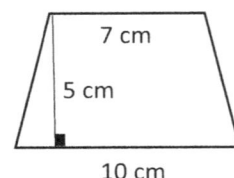

5) _____

6) _____

7) _____

8) _____

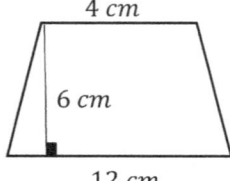

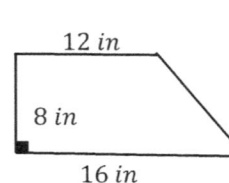

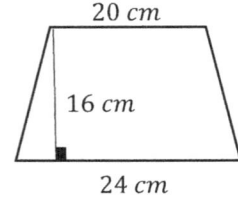

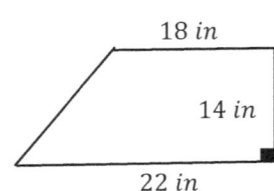

✏️ *Solve.*

9) A trapezoid has an area of 80 cm^2 and its height is 8 cm and one base is 12 cm. What is the other base length? _____

10) If a trapezoid has an area of 120 ft^2 and the lengths of the bases are 14 ft and 16 ft, find the height. _____

11) If a trapezoid has an area of 160 m^2 and its height is 10 m and one base is 14 m, find the other base length. _____

12) The area of a trapezoid is 504 ft^2 and its height is 24 ft. If one base of the trapezoid is 14 ft, what is the other base length? _____

ISEE Middle Level Math Workbook 2020 - 2021

Rectangular Prisms

✎ **Find the volume of each Rectangular Prism.**

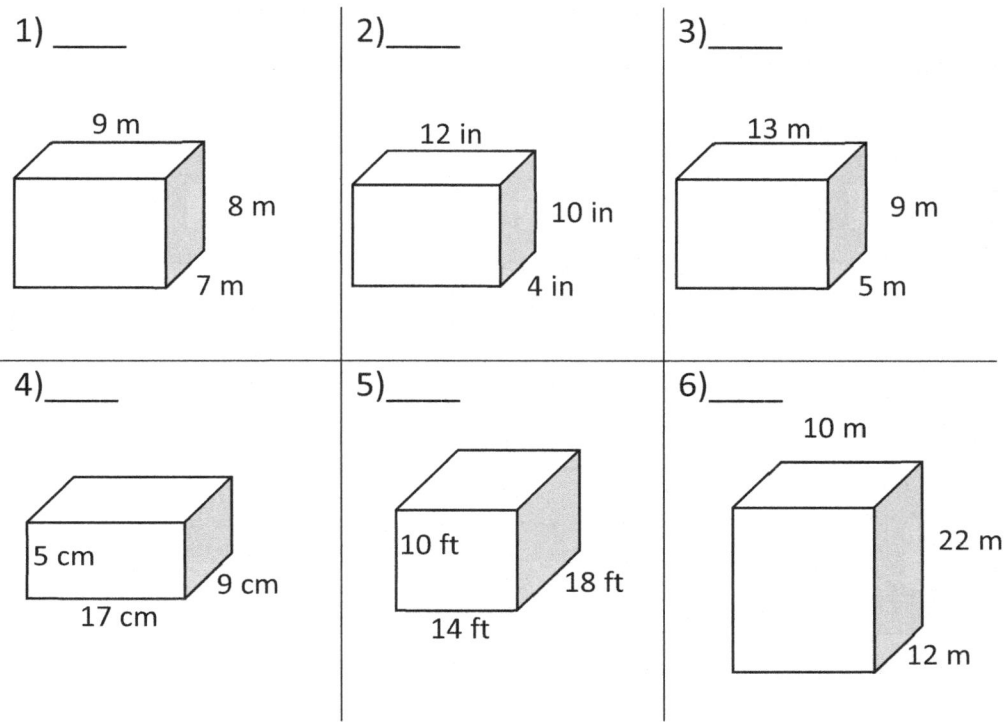

1) ____
9 m
8 m
7 m

2) ____
12 in
10 in
4 in

3) ____
13 m
9 m
5 m

4) ____
5 cm
9 cm
17 cm

5) ____
10 ft
18 ft
14 ft

6) ____
10 m
22 m
12 m

✎ **Find the surface area of each Rectangular Prism.**

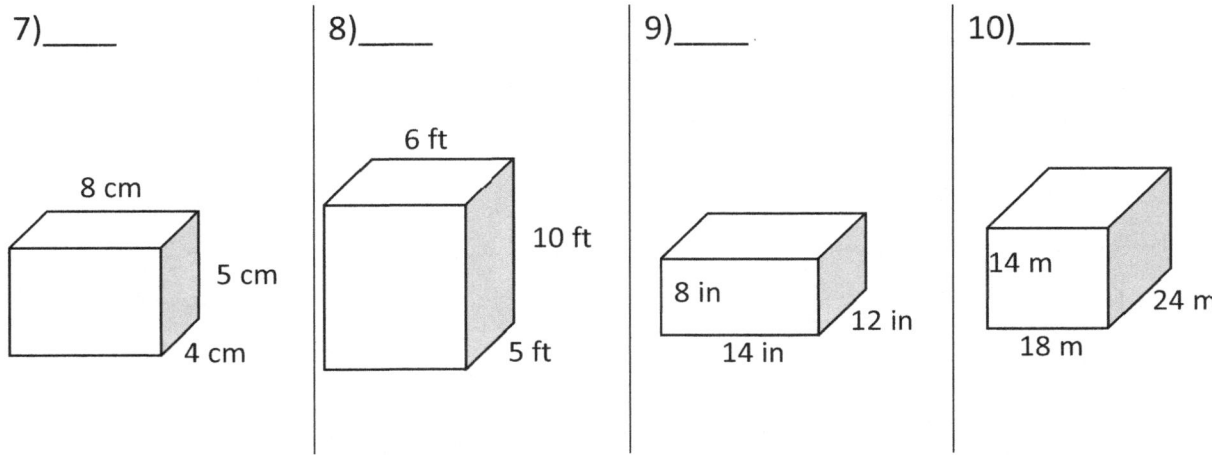

7) ____
8 cm
5 cm
4 cm

8) ____
6 ft
10 ft
5 ft

9) ____
8 in
12 in
14 in

10) ____
14 m
24 m
18 m

Cylinder

✒️ *Find the volume of each Cylinder.* ($\pi = 3.14$)

1) _____
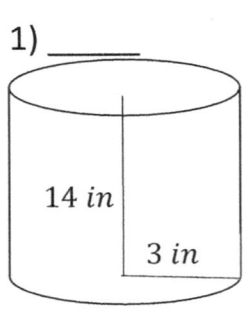
14 in
3 in

2) _____
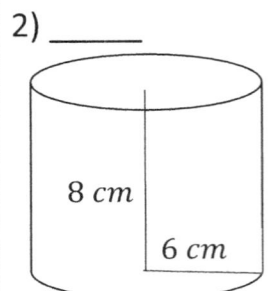
8 cm
6 cm

3) _____
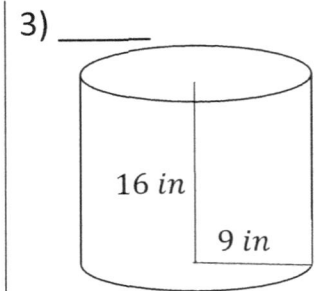
16 in
9 in

4) _____
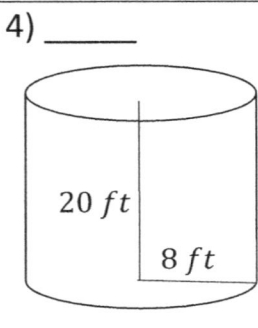
20 ft
8 ft

5) _____
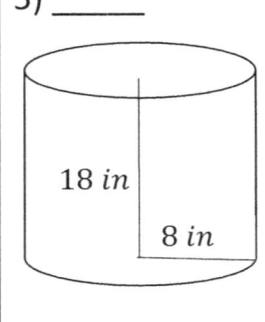
18 in
8 in

6) _____
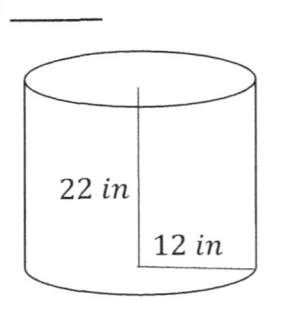
22 in
12 in

✒️ *Find the surface area of each Cylinder.* ($\pi = 3.14$)

7) _____
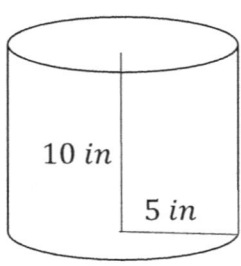
10 in
5 in

8) _____
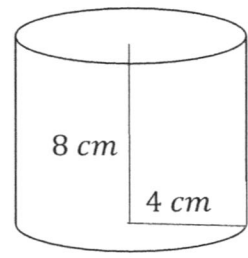
8 cm
4 cm

9) _____
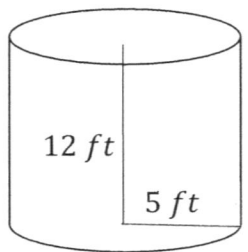
12 ft
5 ft

10) _____
12 m
4 m

ISEE Middle Level Math Workbook 2020 - 2021

Answers – Chapter 9

The Pythagorean Theorem

1) *yes*
2) *yes*
3) *no*
4) *yes*
5) *no*
6) *no*
7) *yes*
8) *yes*
9) 51
10) 12
11) 6
12) 34
13) 26
14) 13
15) 30
16) 52

Triangles

1) 15°
2) 45°
3) 55°
4) 55°
5) 45°
6) 46°
7) 52°
8) 71°
9) 40
10) 56
11) 72 cm^2
12) 42 in^2

Polygons

1) 20 cm
2) 44 m
3) 60 cm
4) 36 m
5) 96 m
6) 56 m
7) 28 cm
8) 120 ft
9) 72 ft
10) 72 in
11) 88 ft
12) 192 in

Circles

1) 43.96 in
2) 75.36 cm
3) 87.92 ft
4) 81.64 m
5) 113.04 cm
6) 94.2 $miles$
7) 119.32 in
8) 138.16 ft
9) 157 m
10) 175.84 m
11) 219.8 in
12) 314 ft

	Radius	Diameter	Circumference	Area
Circle 1	2 inches	4 inches	12.56 inches	12.56 square inches
Circle 2	4 meters	8 meters	25.12 meters	50.24 square meters
Circle 3	6 ft	12 ft	37.68	113.04 square ft
Circle 4	8 miles	16 miles	50.24 miles	200.96 square miles
Circle 5	4.5 km	9 km	28.26 km	63.585 square km
Circle 6	7 cm	14 cm	43.96 cm	153.86 square cm
Circle 7	5 feet	10 feet	31.4 feet	78.5 square feet
Circle 8	14 m	28 m	87.92 m	615.44 square meters
Circle 9	13 in	26 in	81.64 inches	530.66 square inches
Circle 10	12 feet	24 feet	75.36 feet	452.16 square feet

Cubes

1) $91.125\ cm^3$
2) $238.328\ cm^3$
3) $857.375\ ft^3$
4) $1,331\ m^3$
5) $2,197\ in^3$
6) $614.125\ m^3$
7) $421.875\ km^3$
8) $274.625\ cm^3$
9) $4,096\ ft^3$

10) $4,913\ cm^3$
11) $27,000\ in^3$
12) $64,000\ km^3$
13) $3,750\ m^2$
14) $9,600\ m^2$
15) $15,000\ ft^2$
16) $29,400\ mm^2$
17) $21,600\ km^2$
18) $38,400\ cm^2$

Trapezoids

1) $104\ cm^2$
2) $160\ m^2$
3) $224\ ft^2$
4) $324\ cm^2$
5) $288\ cm^2$
6) $414\ in^2$

7) $448\ cm^2$
8) $528\ in^2$
9) $8\ cm$
10) $8\ ft$
11) $18\ m$
12) $28\ ft$

Rectangular Prisms

1) $504\ m^3$
2) $480\ in^3$
3) $585\ m^3$
4) $765\ cm^3$
5) $2,520\ ft^3$

6) $2,640\ m^3$
7) $184\ cm^2$
8) $280\ ft^2$
9) $752\ in^2$
10) $2,040\ m^2$

Cylinder

1) $395.64\ in^3$
2) $904.32\ cm^3$
3) $4,069.44\ in^3$
4) $4,019.2\ ft^3$
5) $3,617.28\ in^3$

6) $9,947.52\ in^3$
7) $471\ in^2$
8) $301.44\ cm^2$
9) $533.8\ ft^2$
10) $401.92\ m^2$

Chapter 10:

Statistics

Math Topics that you'll learn in this Chapter:

- ✓ Mean, Median, Mode, and Range of the Given Data

- ✓ Pie Graph

- ✓ Probability Problems

- ✓ Permutations and Combinations

Mean, Median, Mode, and Range of the Given Data

✍ *Find the values of the Given Data.*

1) $6, 12, 1, 1, 5$

Mode: _____   Range: _____

Mean: _____   Median: _____

2) $5, 8, 3, 7, 4, 3$

Mode: _____   Range: _____

Mean: _____   Median: _____

3) $12, 5, 8, 7, 8$

Mode: _____   Range: _____

Mean: _____   Median: _____

4) $8, 4, 10, 7, 3, 4$

Mode: _____   Range: _____

Mean: _____   Median: _____

5) $9, 7, 10, 5, 7, 4, 14$

Mode: _____   Range: _____

Mean: _____   Median: _____

6) $8, 1, 6, 6, 9, 2, 17$

Mode: _____   Range: _____

Mean: _____   Median: _____

7) $12, 6, 1, 7, 9, 7, 8, 14$

Mode: _____   Range: _____

Mean: _____   Median: _____

8) $10, 14, 5, 4, 11, 6, 13$

Mode: _____   Range: _____

Mean: _____   Median: _____

9) $16, 15, 15, 16, 13, 14, 23$

Mode: _____   Range: _____

Mean: _____   Median: _____

10) $16, 15, 12, 8, 4, 9, 8, 16$

Mode: _____   Range: _____

Mean: _____   Median: _____

Pie Graph

✎ *The circle graph below shows all Wilson's expenses for last month. Wilson spent $200 on his bills last month.*

Answer following questions based on the Pie graph.

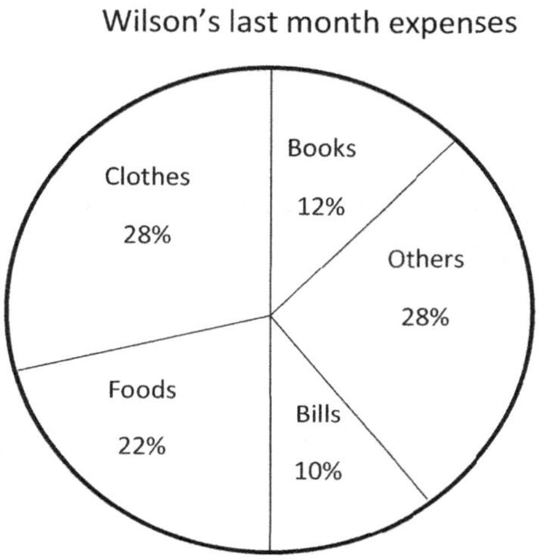

Wilson's last month expenses

1) How much was Wilson's total expenses last month? _____

2) How much did Wilson spend on his clothes last month? _____

3) How much did Wilson spend for foods last month? _____

4) How much did Wilson spend on his books last month? _____

5) What fraction is Wilson's expenses for his bills and clothes out of his total

expenses last month? _____

Probability Problems

1) If there are 10 red balls and 20 blue balls in a basket, what is the probability that Oliver will pick out a red ball from the basket? _____

Gender	Under 45	45 or older	total
Male	12	6	18
Female	5	7	12
Total	17	13	30

2) The table above shows the distribution of age and gender for 30 employees in a company. If one employee is selected at random, what is the probability that the employee selected be either a female under age 45 or a male age 45 or older? _____

3) A number is chosen at random from 1 to 18. Find the probability of not selecting a composite number. (A composite number is a number that is divisible by itself, 1 and at least one other whole number) _____

4) There are 6 blue marbles, 8 red marbles, and 5 yellow marbles in a box. If Ava randomly selects a marble from the box, what is the probability of selecting a red or yellow marble? _____

5) A bag contains 19 balls: three green, five black, eight blue, a brown, a red and one white. If 18 balls are removed from the bag at random, what is the probability that a brown ball has been removed? _____

6) There are only red and blue marbles in a box. The probability of choosing a red marble in the box at random is one fourth. If there are 132 blue marbles, how many marbles are in the box? _____

ISEE Middle Level Math Workbook 2020 - 2021

Permutations and Combinations

✎ *Calculate the value of each.*

1) $5! =$ _____

2) $6! =$ _____

3) $8! =$ _____

4) $5! + 6! =$ _____

5) $8! + 3! =$ _____

6) $6! + 7! =$ _____

7) $8! + 4! =$ _____

8) $9! - 3! =$ _____

✎ *Solve each word problems.*

9) Sophia is baking cookies. She uses milk, flour and eggs. How many different orders of ingredients can she try? _____

10) William is planning for his vacation. He wants to go to restaurant, watch a movie, go to the beach, and play basketball. How many different ways of ordering are there for him? _____

11) How many 7-digit numbers can be named using the digits 1, 2, 3, 4, 5, 6 and 7 without repetition? _____

12) In how many ways can 9 boys be arranged in a straight line? _____

13) In how many ways can 10 athletes be arranged in a straight line? _____

14) A professor is going to arrange her 7 students in a straight line. In how many ways can she do this? _____

15) How many code symbols can be formed with the letters for the word BLACK? _____

16) In how many ways a team of 7 basketball players can choose a captain and co-captain? _____

Answers – Chapter 10

Mean, Median, Mode, and Range of the Given Data

1) Mode: 1   Range: 11   Mean: 5   Median: 5
2) Mode: 3   Range: 5   Mean: 5   Median: 4.5
3) Mode: 8   Range: 7   Mean: 8   Median: 8
4) Mode: 4   Range: 7   Mean: 6   Median: 5.5
5) Mode: 7   Range: 10   Mean: 8   Median: 7
6) Mode: 6   Range: 16   Mean: 7   Median: 6
7) Mode: 7   Range: 13   Mean: 8   Median: 7.5
8) Mode: *no mode* Range: 10   Mean: 9   Median: 10
9) Mode: 15 *and* 16 Range: 10   Mean: 16   Median: 15
10) Mode: 8 *and* 16 Range: 12   Mean: 11   Median: 10.5

Pie Graph

1) $2,000
2) $560
3) $440
4) $240
5) $\frac{19}{50}$

Probability Problems

1) $\frac{1}{3}$
2) $\frac{11}{30}$
3) $\frac{7}{18}$
4) $\frac{13}{19}$
5) $\frac{18}{19}$
6) 176

Permutations and Combinations

1) 120
2) 720
3) 40,320
4) 840
5) 40,326
6) 5,760
7) 40,344
8) 362,874
9) *6*
10) *24*
11) 5,040
12) 362,880
13) 3,628,800
14) 5,040
15) 120
16) 42

ISEE Middle Level Test Review

The Independent School Entrance Exam (ISEE) is an admission test developed by the Educational Records Bureau for its member schools as part of their admission process.

ISEE Middle Level tests use a multiple-choice format and contain two Mathematics sections:

Quantitative Reasoning

There are 37 questions in the Quantitative Reasoning section and students have 35 minutes to answer the questions. This section contains word problems and quantitative comparisons. The word problems require either no calculation or simple calculation. The quantitative comparison items present two quantities, (A) and (B), and the student needs to select one of the following four answer choices:

(A) The quantity in Column A is greater.

(B) The quantity in Column B is greater.

(C) The two quantities are equal.

(D) The relationship cannot be determined from the information given.

Mathematics Achievement

There are 47 questions in the Mathematics Achievement section and students have 40 minutes to answer the questions. Mathematics Achievement measures students' knowledge of Mathematics requiring one or more steps in calculating the answer.

In this book, there are two complete ISEE Middle Level Quantitative Reasoning and Mathematics Achievement practice tests. Take these tests to see what score you'll be able to receive on a real ISEE Lower Level test.

Good luck!

Time to refine your skill with a practice examination

Take practice ISEE Middle Level Math Tests to simulate the test day experience. After you've finished, score your tests using the answer keys.

Before You Start

- You'll need a pencil and a timer to take the test.
- After you've finished the test, review the answer key to see where you went wrong.
- Use the answer sheet provided to record your answers. (You can cut it out or photocopy it)
- Students receive 1 point for every correct answer. There is no penalty for wrong or skipped questions.

Calculators are NOT permitted for the ISEE Middle Level Test

Good Luck!

ISEE Middle Level Math Practice Test 1

2020 - 2021

Two Parts

Total number of questions: 84

Part 1 (Quantitative Reasoning): 37 questions

Part 2 (Mathematics Achievement): 47 questions

Total time for two parts: 75 Minutes

ISEE Middle Level Practice Test Answer Sheets

Remove (or photocopy) this answer sheet and use it to complete the practice test.

ISEE Middle Level Practice Test 1

Quantitative Reasoning

1. Ⓐ Ⓑ Ⓒ Ⓓ
2. Ⓐ Ⓑ Ⓒ Ⓓ
3. Ⓐ Ⓑ Ⓒ Ⓓ
4. Ⓐ Ⓑ Ⓒ Ⓓ
5. Ⓐ Ⓑ Ⓒ Ⓓ
6. Ⓐ Ⓑ Ⓒ Ⓓ
7. Ⓐ Ⓑ Ⓒ Ⓓ
8. Ⓐ Ⓑ Ⓒ Ⓓ
9. Ⓐ Ⓑ Ⓒ Ⓓ
10. Ⓐ Ⓑ Ⓒ Ⓓ
11. Ⓐ Ⓑ Ⓒ Ⓓ
12. Ⓐ Ⓑ Ⓒ Ⓓ
13. Ⓐ Ⓑ Ⓒ Ⓓ
14. Ⓐ Ⓑ Ⓒ Ⓓ
15. Ⓐ Ⓑ Ⓒ Ⓓ
16. Ⓐ Ⓑ Ⓒ Ⓓ
17. Ⓐ Ⓑ Ⓒ Ⓓ
18. Ⓐ Ⓑ Ⓒ Ⓓ
19. Ⓐ Ⓑ Ⓒ Ⓓ
20. Ⓐ Ⓑ Ⓒ Ⓓ
21. Ⓐ Ⓑ Ⓒ Ⓓ
22. Ⓐ Ⓑ Ⓒ Ⓓ
23. Ⓐ Ⓑ Ⓒ Ⓓ
24. Ⓐ Ⓑ Ⓒ Ⓓ

25. Ⓐ Ⓑ Ⓒ Ⓓ
26. Ⓐ Ⓑ Ⓒ Ⓓ
27. Ⓐ Ⓑ Ⓒ Ⓓ
28. Ⓐ Ⓑ Ⓒ Ⓓ
29. Ⓐ Ⓑ Ⓒ Ⓓ
30. Ⓐ Ⓑ Ⓒ Ⓓ
31. Ⓐ Ⓑ Ⓒ Ⓓ
32. Ⓐ Ⓑ Ⓒ Ⓓ
33. Ⓐ Ⓑ Ⓒ Ⓓ
34. Ⓐ Ⓑ Ⓒ Ⓓ
35. Ⓐ Ⓑ Ⓒ Ⓓ
36. Ⓐ Ⓑ Ⓒ Ⓓ
37. Ⓐ Ⓑ Ⓒ Ⓓ

Mathematics Achievement

1. Ⓐ Ⓑ Ⓒ Ⓓ
2. Ⓐ Ⓑ Ⓒ Ⓓ
3. Ⓐ Ⓑ Ⓒ Ⓓ
4. Ⓐ Ⓑ Ⓒ Ⓓ
5. Ⓐ Ⓑ Ⓒ Ⓓ
6. Ⓐ Ⓑ Ⓒ Ⓓ
7. Ⓐ Ⓑ Ⓒ Ⓓ
8. Ⓐ Ⓑ Ⓒ Ⓓ
9. Ⓐ Ⓑ Ⓒ Ⓓ
10. Ⓐ Ⓑ Ⓒ Ⓓ
11. Ⓐ Ⓑ Ⓒ Ⓓ
12. Ⓐ Ⓑ Ⓒ Ⓓ
13. Ⓐ Ⓑ Ⓒ Ⓓ
14. Ⓐ Ⓑ Ⓒ Ⓓ
15. Ⓐ Ⓑ Ⓒ Ⓓ
16. Ⓐ Ⓑ Ⓒ Ⓓ
17. Ⓐ Ⓑ Ⓒ Ⓓ
18. Ⓐ Ⓑ Ⓒ Ⓓ
19. Ⓐ Ⓑ Ⓒ Ⓓ
20. Ⓐ Ⓑ Ⓒ Ⓓ
21. Ⓐ Ⓑ Ⓒ Ⓓ
22. Ⓐ Ⓑ Ⓒ Ⓓ
23. Ⓐ Ⓑ Ⓒ Ⓓ
24. Ⓐ Ⓑ Ⓒ Ⓓ

25. Ⓐ Ⓑ Ⓒ Ⓓ
26. Ⓐ Ⓑ Ⓒ Ⓓ
27. Ⓐ Ⓑ Ⓒ Ⓓ
28. Ⓐ Ⓑ Ⓒ Ⓓ
29. Ⓐ Ⓑ Ⓒ Ⓓ
30. Ⓐ Ⓑ Ⓒ Ⓓ
31. Ⓐ Ⓑ Ⓒ Ⓓ
32. Ⓐ Ⓑ Ⓒ Ⓓ
33. Ⓐ Ⓑ Ⓒ Ⓓ
34. Ⓐ Ⓑ Ⓒ Ⓓ
35. Ⓐ Ⓑ Ⓒ Ⓓ
36. Ⓐ Ⓑ Ⓒ Ⓓ
37. Ⓐ Ⓑ Ⓒ Ⓓ
38. Ⓐ Ⓑ Ⓒ Ⓓ
39. Ⓐ Ⓑ Ⓒ Ⓓ
40. Ⓐ Ⓑ Ⓒ Ⓓ
41. Ⓐ Ⓑ Ⓒ Ⓓ
42. Ⓐ Ⓑ Ⓒ Ⓓ
43. Ⓐ Ⓑ Ⓒ Ⓓ
44. Ⓐ Ⓑ Ⓒ Ⓓ
45. Ⓐ Ⓑ Ⓒ Ⓓ
46. Ⓐ Ⓑ Ⓒ Ⓓ
47. Ⓐ Ⓑ Ⓒ Ⓓ

ISEE Middle Level Math Practice Test 1

Section 1

37 questions

Total time for this section: 35 Minutes

You may NOT use a calculator for this test

1) What is the value of x in the following equation?
$$\frac{7^x}{7} = 343$$

A. 4
B. 5
C. 8
D. 12

2) In following shape y equals to?

A. 120°
B. 30°
C. 25.5°
D. 20°

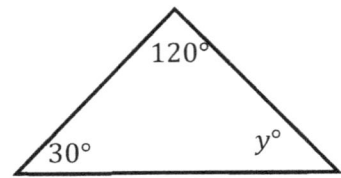

3) Which of the following shows the numbers in increasing order?
A. $\frac{1}{3}, \frac{8}{12}, \frac{4}{7}, \frac{3}{4}$
B. $\frac{1}{3}, \frac{4}{7}, \frac{8}{12}, \frac{3}{4}$
C. $\frac{4}{7}, \frac{3}{4}, \frac{8}{12}, \frac{1}{3}$
D. $\frac{8}{12}, \frac{3}{4}, \frac{4}{7}, \frac{1}{3}$

4) If an object travels at $0.4\ cm$ per second, how many meters does it travel in 5 hours?
A. $88.2\ m$
B. $76.4\ m$
C. $72\ m$
D. $43.2\ m$

5) If the ratio of home fans to visiting fans in a crowd is $3:2$ and all 24,000 seats in a stadium are filled, how many visiting fans are in attendance?
A. 96,000
B. 9,600
C. 960
D. 96

6) What's the approximate circumference of a circle that has a diameter of $17m$?
A. $53.38\ m$
B. $71.9\ m$
C. $97.25\ m$
D. $100\ m$

7) What is the lowest common multiple of 24 and 36?
A. 48
B. 72
C. 108
D. 864

8) What is the area of the shaded region? (one fourth of the circle is shaded) (Diameter = 8)
A. 4π
B. 6π
C. 8π
D. 9π

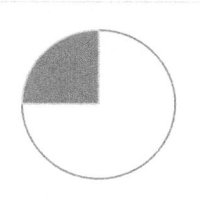

9) An item in the store originally priced at $200 was marked down 30%. What is the final sale price of the item?
A. $240
B. $204
C. $200
D. $140

10) A shirt costing $300 is discounted 15%. After a month, the shirt is discounted another 25%. Which of the following expressions can be used to find the selling price of the shirt?
A. $(300)(0.70)$
B. $(300) - 300(0.30)$
C. $(300)(0.15) - (400)(0.15)$
D. $(300)(0.85)(0.75)$

11) If a car has 70-liter petrol and after one hour driving the car use 5-liter petrol, how much petrol remaining after x-hours?
A. $5x - 70$
B. $70 + 5x$
C. $70 - 5x$
D. $70 - x$

12) Solve for x: $4 + x + 8\left(\frac{x}{4}\right) = 2x + 12$
A. 8
B. 5.5
C. 4
D. 4.5

13) The area of the trapezoid below is 136. What is the value of x?

A. 7
B. 8
C. 10
D. 11

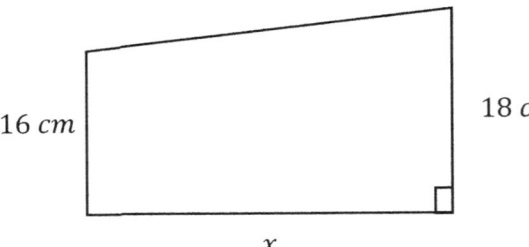

16 cm

18 cm

x

14) Find $\frac{1}{3}$ of $\frac{1}{2}$ of $\frac{3}{5}$ of 280?

A. 28
B. 30
C. 31
D. 2

15) If $x \leq a$ is the solution of $6 + 3x \leq 21$, what is the value of a?

A. $21x$
B. 5
C. -5
D. $15x$

16) 7 liters of water are poured into an aquarium that's $25cm$ long, $5cm$ wide, and $70cm$ high. How many cm will the water level in the aquarium rise due to this added water? ($1\ liter\ of\ water\ =\ 1,000\ cm^3$)

A. 80
B. 56
C. 49
D. 10

17) If $4f + 4g = 4x - 2y$ and $g = 2y - 6x$, what is $2f$?

A. $5x + y$
B. $14x + 3y$
C. $14x - 5y$
D. $y - 3x$

18) What is the value of $\dfrac{-\frac{13}{3} \times \frac{4}{5}}{\frac{10}{30}}$?

A. -10.4
B. 10.4
C. $-\frac{1}{9}$
D. $\frac{1}{9}$

19) What is the perimeter of the following parallelogram?

A. 54
B. 44
C. 24
D. 17

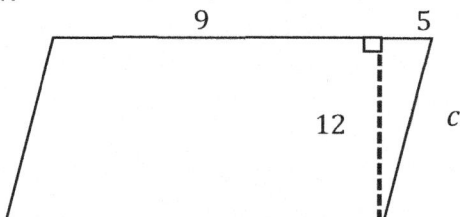

20) In a bundle of 50 fruits, 6 are apples and the rest are bananas. What percent of the bundle is composed of apples?

A. 30%
B. 25%
C. 12%
D. 10%

21) 3 less than twice a positive integer is 71. What is the integer?

A. 37
B. 40
C. 42
D. 44

22) If Joe was making $8.50 per hour and got a raise of $0.35 per hour, approximately what percentage increase was the raise?

A. 2%
B. 2.67%
C. 3.33%
D. 4.00%

23) Which is the equivalent temperature of $140°F$ in Celsius? ($C = Celsius$)

$$C = \frac{5}{9}(F - 32)$$

A. 32
B. 38.5
C. 50
D. 60

24) The average of $14, 16, 21$ and x is 20. What is the value of x?

A. 9
B. 15
C. 18
D. 29

25) What is the value of mode and median in the following set of numbers?

2 ,3, 3, 6, 5, 5, 4, 4, 6, 2, 2

A. Mode: 2 Median:4
B. Mode:2, Median:4
C. Mode:2, 3 Median:5
D. Mode: 3 Median:4

Quantitative Comparisons

Direction: Questions 26 to 37 are Quantitative Comparisons Questions. Using the information provided in each question, compare the quantity in column A to the quantity in Column B. Choose on your answer sheet grid

- A if the quantity in Column A is greater
- B if the quantity in Column B is greater
- C if the two quantities are equal
- D if the relationship cannot be determined from the information given

26)

Column A	Column B
$\dfrac{\sqrt{64-48}}{\sqrt{25-9}}$	$\dfrac{(7-4)}{(8-3)}$

27) $2x^5 - 9 = 477$
$\dfrac{1}{3} - \dfrac{y}{5} = -\dfrac{7}{15}$

Quantity A	Quantity B
x	y

28) The sum of 3 consecutive integers is -45.

Column A	Column B
The largest of these integers	-16

29)

Column A	Column B
$4^2 - 2^4$	$2^4 - 4^2$

30) A computer costs $250.

Column A	Column B
A sales tax at 8% of the computer cost	$20

31) A

Column A	Column B
The slope of the line $4x + 2y = 7$	The slope of the line that passes through points $(2, 5)$ and $(3, 3)$

32)

Quantity A	Quantity B
The least prime factor of 55	The least prime factor of 210

33)

Column A	Column B
$\sqrt{144 - 81}$	$\sqrt{144} - \sqrt{81}$

34) 6 percent of x is equal to 5 percent of y, where x and y are positive numbers.

Quantity A	Quantity B
x	y

35)

Quantity A	Quantity B
$(-5)^4$	5^4

36)

Quantity A	Quantity B
$(1.88)^4(1.88)^8$	$(1.88)^{12}$

37) x is a positive number.

Quantity A	Quantity B
x^{10}	x^{20}

STOP

ISEE Middle Level Math

Practice Test 1

Section 2

47 questions

Total time for this section: 40 Minutes

You may NOT use a calculator for this test.

1) Which of the following is not synonym for 20^2?
A. 20 cubed
B. 20 squared
C. The square of 20
D. 20 to the second power

2) What is the value of x in the following equation?
$$3^x + 28 = 55$$
A. 3
B. 4
C. 5
D. 6

3) If angles A and B are angles of a parallelogram, what is the sum of the measures of the two angles?
A. 360 degrees
B. 180 degrees
C. 90 degrees
D. Cannot be determined

4) If x = lowest common multiple of 10 and 35, then $\frac{x}{5} + 2$ equal to?
A. 70
B. 58
C. 46
D. 16

5) If the area of the following trapezoid is 30, what is the perimeter of the trapezoid?
A. 25
B. 28
C. 45
D. 55

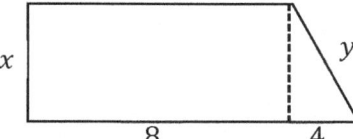

6) A swing moves from one extreme point (point A) to the opposite extreme point (point B) in 20 seconds. How long does it take that the swing moves 5 times from point A to point B and returns to point A?
A. 400 seconds
B. 200 seconds
C. 150 seconds
D. 100 seconds

7) There are 2 cars moving in the same direction on a road. A red car is 12 km ahead of a blue car. If the speed of the red car is 40 $km\ per\ hour$ and the speed of the blue car is $1\frac{2}{5}$ of the red car, how many minutes will it take the blue car to catch the red car?

A. 8.5
B. 15
C. 30
D. 45

8) In two successive years, the population of a town is increased by 10% and 25%. What percent of the population is increased after two years?

A. 25%
B. 35%
C. 36%
D. 37%

9) $5 + 8 \times (-2) - [4 + 22 \times 5] \div 6\ =\ ?$

A. 120
B. 88
C. −30
D. −20

10) In 1989, the average worker's income increased $2,500 per year starting from $26,000 annual salary. Which equation represents income greater than average?
(I = income, x = number of years after 1989)

A. $I > 2{,}500x + 26{,}000$
B. $I > -2{,}500x + 26{,}000$
C. $I < -2{,}500x + 26{,}000$
D. $I < 2{,}500x - 26{,}000$

11) Which of the following angles is obtuse?

A. 10 degrees
B. 30 degrees
C. 189 degrees
D. 120 degrees

12) Mr. Jones saves $2,500 out of his monthly family income of $65,000. What fractional part of his income does he save?

A. $\frac{1}{26}$

B. $\frac{1}{11}$

C. $\frac{3}{26}$

D. $\frac{2}{15}$

13) Anita's trick–or–treat bag contains 13 pieces of chocolate, 19 suckers, 19 pieces of gum, 25 pieces of licorice. If she randomly pulls a piece of candy from her bag, what is the probability of her pulling out a piece of sucker

A. $\frac{1}{3}$

B. $\frac{1}{4}$

C. $\frac{1}{5}$

D. $\frac{1}{19}$

14) 110 is equal to?

A. $20 - (4 \times 10) + (6 \times 30)$

B. $\left(\frac{11}{8} \times 72\right) + \left(\frac{125}{5}\right)$

C. $\left(\left(\frac{30}{4} + \frac{13}{2}\right) \times 7\right) - \frac{11}{2} + \frac{110}{4}$

D. $(2 \times 10) + (50 \times 1.5) + 15$

15) What is the difference in area between a $8\ cm$ by $4\ cm$ rectangle and a circle with diameter of $8\ cm$? ($\pi = 3$)

A. $49\ cm$

B. $40\ cm$

C. $39\ cm$

D. $16\ cm$

16) Solve the following equation?
$$(x^2 + 2x + 1) = 64$$

A. $-9, -7$

B. $-9, 7$

C. -9

D. 7

17) What is ratio of perimeter of figure A to area of figure B?

A. $\frac{1}{3}$

B. $\frac{3}{8}$

C. 3

D. 5

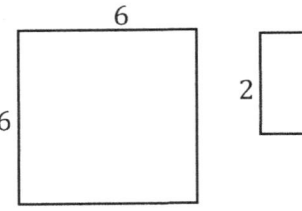

Fig. A

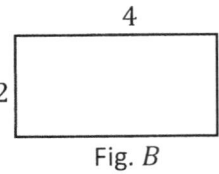

Fig. B

18) $\frac{18 \times 21}{5}$ is closest estimate to?

A. 75.6

B. 68.7

C. 50.6

D. 40.7

19) How many possible outfit combinations come from four shirts, two slacks, and five ties?

A. 60

B. 40

C. 16

D. 10

20) When a number is multiplied to itself and added by 9, the result is 25. What is the value of the number?

A. 4 or -4

B. 5 or -5

C. 4

D. 5

21) If you invest \$2,000 at an annual rate of 8%, how much interest will you earn after one year?

A. 16,000

B. 16,00

C. 380

D. 160

22) What is the value of x in the equation: $\frac{x}{4} + \frac{5}{4} = 5$?

A. 15

B. 10

C. 8

D. 5

23) If $y = 5ab + 3b^3$, what is y when $a = 3$ and $b = 2$?
A. 64
B. 65
C. 55
D. 54

24) What is the absolute value of the quantity six minus ten?
A. -4
B. 10
C. -10
D. 4

25) Which of the following angles can represent the three angles of an equilateral triangle?
A. $45°, 90°, 45°$
B. $50°, 50°, 80°$
C. $60°, 60°, 60°$
D. $55°, 35°, 90°$

26) In the following equation, what is the value of $x + y$?
$$9x - 10 = 5\left(\frac{4}{5}x - y\right) + 5$$

A. 15
B. -15
C. 3
D. -3

27) How many tiles of $3\ cm^2$ is needed to cover a floor of dimension $7\ cm$ by $27\ cm$?
A. 12
B. 38
C. 63
D. 66

28) Two-kilogram apple and three-kilograms orange cost $21. If the price of one-kilogram of apple is twice the price of one-kilogram of orange, how much does one-kilogram apple cost?
A. $8
B. $6
C. $4
D. $1

29) Which is **NOT** a prime number?
A. 181
B. 151
C. 131
D. 122

30) Each of the x students in a team may invite up to 6 friends to a party. What is the maximum number of students and guests who might attend the party?
A. $6x + 6$
B. $6x$
C. $x + 6$
D. $7x$

31) Calculate the approximate circumference of the following circle. (the diameter is 10)

A. 1,267
B. 314
C. 31
D. 10

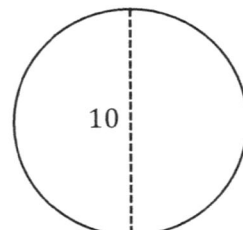

32) 280 $minutes$=...?
A. 5 $Hours$
B. 4.6 $Hours$
C. 3.5 $Hours$
D. 0.5 $Hours$

33) There are three boxes, a red box, a blue box, and a yellow box. If the weight of the red box is 60 kg and it is 80% of the weight of the blue box, and the weight of the blue box is 120% of the weight of the yellow box, what is the weight of all boxes?
A. 197.5 kg
B. 210.5 kg
C. 280 kg
D. 320 kg

34) In the figure below, line A is parallel to line B. What is the value of angle x?

A. 35 degree
B. 40 degree
C. 100 degree
D. 140 degree

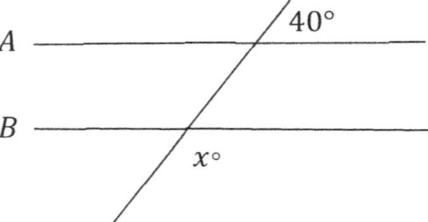

35) Jim drove 350 miles and it took him approximately 9 hours. How many miles per hour was his average speed?

A. about 34.7 miles per hour
B. about 38.8 miles per hour
C. about 48.5 miles per hour
D. about 49.5 miles per hour

36) Three people go to a restaurant. Their bill comes to $58.00. They decided to split the cost. One person pays $7.5, the next person pays 2 times that amount. How much will the third person have to pay?

A. $34.50
B. $35.50
C. $41.00
D. $45.00

37) $\left(\left((-15) + 40\right) \times \frac{1}{5}\right) + (-15)$?

A. 5
B. 10
C. −5
D. −10

38) What is 13,8210 in scientific notation?

A. 138.21×10^3
B. 13.821×10^4
C. 0.13821×10^6
D. 1.3821×10^5

39) What is the value of $(10 - 6)!$?

A. 20
B. 24
C. 26
D. 28

40) What is the perimeter of the below right triangle?

A. 20
B. 18
C. 15
D. 12

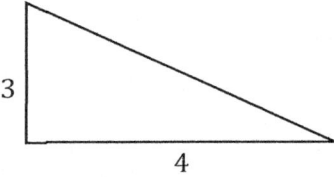

41) If 150% of a number is equal to 30% of 80, then what is the number?
   A. 16
   B. 15.5
   C. 14.66
   D. 12.25

42) In a department of a company, the ratio of employees with Bachelor's Degree to employees with high school Diploma is 1 to 4. If there are 24 employees with Bachelor's Degree in this department, how many employees with High School Diploma should be moved to other departments to change the ratio of the number of employees with Bachelor's Degree to the number of employees with High School Diploma to 3 to 4 in this department?
   A. 64
   B. 54
   C. 10
   D. 12

43) What is x in the following right triangle?

   A. $\sqrt{399}$
   B. 15
   C. 20
   D. $\sqrt{402}$

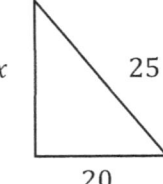

44) The average weight of 20 girls in a class is 65 kg and the average weight of 35 boys in the same class is 72 kg. What is the average weight of all the 55 students in that class?
   A. 61.28
   B. 61.68
   C. 62.90
   D. 69.45

45) An angle is equal to one ninth of its supplement. What is the measure of that angle?
   A. 18
   B. 30
   C. 45
   D. 60

46) What is the difference of smallest 5–digit number and biggest 5–digit number?
   A. 66,666
   B. 67,899
   C. 88,888
   D. 89,999

47) John traveled 140 km in 5 hours and Alice traveled 210 km in 3 hours. What is the ratio of the average speed of John to average speed of Alice?

A. 2 : 5
B. 5 : 2
C. 5 : 9
D. 5 : 6

IF YOU FINISH BEFORE TIME IS CALLED, YOU MAY CHECK YOUR WORK ON THIS SECTION. **STOP**

ISEE Middle Level Math Practice Test 2

2020 - 2021

Two Parts

Total number of questions: 84

Part 1 (Quantitative Reasoning): 37 questions

Part 2 (Mathematics Achievement): 47 questions

Total time for two parts: 75 Minutes

ISEE Middle Level Practice Test Answer Sheets

Remove (or photocopy) this answer sheet and use it to complete the practice test.

ISEE Middle Level Practice Test 2

Quantitative Reasoning

1. Ⓐ Ⓑ Ⓒ Ⓓ
2. Ⓐ Ⓑ Ⓒ Ⓓ
3. Ⓐ Ⓑ Ⓒ Ⓓ
4. Ⓐ Ⓑ Ⓒ Ⓓ
5. Ⓐ Ⓑ Ⓒ Ⓓ
6. Ⓐ Ⓑ Ⓒ Ⓓ
7. Ⓐ Ⓑ Ⓒ Ⓓ
8. Ⓐ Ⓑ Ⓒ Ⓓ
9. Ⓐ Ⓑ Ⓒ Ⓓ
10. Ⓐ Ⓑ Ⓒ Ⓓ
11. Ⓐ Ⓑ Ⓒ Ⓓ
12. Ⓐ Ⓑ Ⓒ Ⓓ
13. Ⓐ Ⓑ Ⓒ Ⓓ
14. Ⓐ Ⓑ Ⓒ Ⓓ
15. Ⓐ Ⓑ Ⓒ Ⓓ
16. Ⓐ Ⓑ Ⓒ Ⓓ
17. Ⓐ Ⓑ Ⓒ Ⓓ
18. Ⓐ Ⓑ Ⓒ Ⓓ
19. Ⓐ Ⓑ Ⓒ Ⓓ
20. Ⓐ Ⓑ Ⓒ Ⓓ
21. Ⓐ Ⓑ Ⓒ Ⓓ
22. Ⓐ Ⓑ Ⓒ Ⓓ
23. Ⓐ Ⓑ Ⓒ Ⓓ
24. Ⓐ Ⓑ Ⓒ Ⓓ

25. Ⓐ Ⓑ Ⓒ Ⓓ
26. Ⓐ Ⓑ Ⓒ Ⓓ
27. Ⓐ Ⓑ Ⓒ Ⓓ
28. Ⓐ Ⓑ Ⓒ Ⓓ
29. Ⓐ Ⓑ Ⓒ Ⓓ
30. Ⓐ Ⓑ Ⓒ Ⓓ
31. Ⓐ Ⓑ Ⓒ Ⓓ
32. Ⓐ Ⓑ Ⓒ Ⓓ
33. Ⓐ Ⓑ Ⓒ Ⓓ
34. Ⓐ Ⓑ Ⓒ Ⓓ
35. Ⓐ Ⓑ Ⓒ Ⓓ
36. Ⓐ Ⓑ Ⓒ Ⓓ
37. Ⓐ Ⓑ Ⓒ Ⓓ

Mathematics Achievement

1. Ⓐ Ⓑ Ⓒ Ⓓ
2. Ⓐ Ⓑ Ⓒ Ⓓ
3. Ⓐ Ⓑ Ⓒ Ⓓ
4. Ⓐ Ⓑ Ⓒ Ⓓ
5. Ⓐ Ⓑ Ⓒ Ⓓ
6. Ⓐ Ⓑ Ⓒ Ⓓ
7. Ⓐ Ⓑ Ⓒ Ⓓ
8. Ⓐ Ⓑ Ⓒ Ⓓ
9. Ⓐ Ⓑ Ⓒ Ⓓ
10. Ⓐ Ⓑ Ⓒ Ⓓ
11. Ⓐ Ⓑ Ⓒ Ⓓ
12. Ⓐ Ⓑ Ⓒ Ⓓ
13. Ⓐ Ⓑ Ⓒ Ⓓ
14. Ⓐ Ⓑ Ⓒ Ⓓ
15. Ⓐ Ⓑ Ⓒ Ⓓ
16. Ⓐ Ⓑ Ⓒ Ⓓ
17. Ⓐ Ⓑ Ⓒ Ⓓ
18. Ⓐ Ⓑ Ⓒ Ⓓ
19. Ⓐ Ⓑ Ⓒ Ⓓ
20. Ⓐ Ⓑ Ⓒ Ⓓ
21. Ⓐ Ⓑ Ⓒ Ⓓ
22. Ⓐ Ⓑ Ⓒ Ⓓ
23. Ⓐ Ⓑ Ⓒ Ⓓ
24. Ⓐ Ⓑ Ⓒ Ⓓ

25. Ⓐ Ⓑ Ⓒ Ⓓ
26. Ⓐ Ⓑ Ⓒ Ⓓ
27. Ⓐ Ⓑ Ⓒ Ⓓ
28. Ⓐ Ⓑ Ⓒ Ⓓ
29. Ⓐ Ⓑ Ⓒ Ⓓ
30. Ⓐ Ⓑ Ⓒ Ⓓ
31. Ⓐ Ⓑ Ⓒ Ⓓ
32. Ⓐ Ⓑ Ⓒ Ⓓ
33. Ⓐ Ⓑ Ⓒ Ⓓ
34. Ⓐ Ⓑ Ⓒ Ⓓ
35. Ⓐ Ⓑ Ⓒ Ⓓ
36. Ⓐ Ⓑ Ⓒ Ⓓ
37. Ⓐ Ⓑ Ⓒ Ⓓ
38. Ⓐ Ⓑ Ⓒ Ⓓ
39. Ⓐ Ⓑ Ⓒ Ⓓ
40. Ⓐ Ⓑ Ⓒ Ⓓ
41. Ⓐ Ⓑ Ⓒ Ⓓ
42. Ⓐ Ⓑ Ⓒ Ⓓ
43. Ⓐ Ⓑ Ⓒ Ⓓ
44. Ⓐ Ⓑ Ⓒ Ⓓ
45. Ⓐ Ⓑ Ⓒ Ⓓ
46. Ⓐ Ⓑ Ⓒ Ⓓ
47. Ⓐ Ⓑ Ⓒ Ⓓ

ISEE Middle Level Math
Practice Test 2

Section 1

37 questions

Total time for this section: 35 Minutes

You may NOT use a calculator for this test.

1) If the ratio of home fans to visiting fans in a crowd is $3:2$ and all 25,000 seats in a stadium are filled, how many visiting fans are in attendance?
A. 100,000
B. 10,000
C. 1,000
D. 100

2) In following shape y equals to?

A. 128.5°
B. 51.5°
C. 48.5°
D. 35°

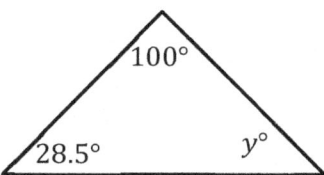

3) Which of the following shows the numbers in increasing order?
A. $\frac{2}{3}, \frac{8}{11}, \frac{5}{7}, \frac{3}{4}$
B. $\frac{2}{3}, \frac{5}{7}, \frac{8}{11}, \frac{3}{4}$
C. $\frac{5}{7}, \frac{3}{4}, \frac{8}{11}, \frac{2}{3}$
D. $\frac{8}{11}, \frac{3}{4}, \frac{5}{7}, \frac{2}{3}$

4) If an object travels at $0.3\ cm$ per second, how many meters does it travel in 4 hours?
A. $88.2\ m$
B. $66.4\ m$
C. $50\ m$
D. $43.2\ m$

5) What is the value of x in the following equation?
$$\frac{3^x}{9} = 243$$
A. 7
B. 5
C. 3
D. 2

6) Ava uses a 40% off coupon when buying a sweater that costs $40. If she also pays 5% sales tax on the purchase, how much does she pay?
A. $26.95
B. $25.20
C. $21.7
D. $14.83

7) An item in the store originally priced at $300 was marked down 20%. What is the final sale price of the item?

A. $240
B. $204
C. $200
D. $195

8) What's the circumference of a circle that has a diameter of $15\ m$? ($\pi = 3.14$)

A. $47.1\ m$
B. $94.29\ m$
C. $150\ m$
D. $225\ m$

9) What is the area of the shaded region? (one forth of the circle is shaded)
Diameter $= 12$

A. $6\,\pi$
B. $7\,\pi$
C. $8\,\pi$
D. $9\,\pi$

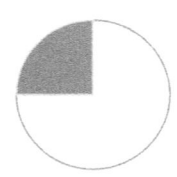

10) If a car has 80-liter petrol and after one hour driving the car use 6-liter petrol, how much petrol remaining after x-hours?

A. $6x - 80$
B. $80 + 6x$
C. $80 - 6x$
D. $80 - x$

11) A shirt costing $200 is discounted 15%. After a month, the shirt is discounted another 15%. Which of the following expressions can be used to find the selling price of the shirt

A. $(200)\,(0.70)$
B. $(200) - 200\,(0.30)$
C. $(200)(0.15) - (200)\,(0.15)$
D. $(200)\,(0.85)\,(0.85)$

12) Find $\frac{1}{3}$ of $\frac{2}{5}$ of $\frac{3}{4}$ of 300?

A. 30
B. 32
C. 35
D. 45

13) If $x \le a$ is the solution of $7 + 2x \le 15$, what is the value of a?
A. $14x$
B. 4
C. -4
D. $15x$

14) The area of the trapezoid below is 132. What is the value of x?

A. 8
B. 9
C. 10
D. 11

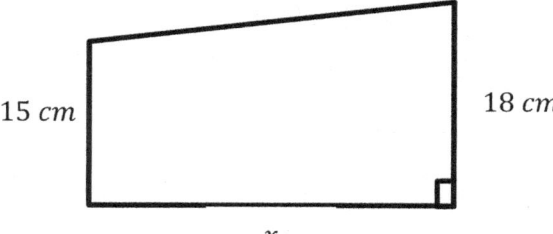

15) Solve for x: $3 + x + 6\left(\frac{x}{2}\right) = 2x + 10$
A. $\frac{13}{6}$
B. $\frac{7}{6}$
C. $\frac{7}{2}$
D. $\frac{13}{2}$

16) 6 liters of water are poured into an aquarium that's $15cm$ long, $5cm$ wide, and $90cm$ high. How many cm will the water level in the aquarium rise due to this added water?
($1 \ liter \ of \ water = 1,000 \ cm^3$)
A. 80
B. 40
C. 20
D. 10

17) If $3f + 2g = 3x + y$ and $g = 2y - 3x$, what is f?
A. $3x + y$
B. $x + 3y$
C. $3x - y$
D. $y - 3x$

18) What is the perimeter of the following parallelogram?
A. 48
B. 34
C. 24
D. 17

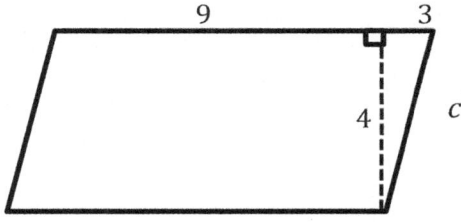

19) In a bundle of 40 fruits, 8 are apples and the rest are bananas. What percent of the bundle is composed of apples?

A. 40%

B. 25%

C. 20%

D. 15%

20) What is the value of $\dfrac{-\frac{11}{2}\times\frac{3}{5}}{\frac{11}{30}}$?

A. -9

B. 9

C. $-\frac{1}{9}$

D. $\frac{1}{9}$

21) The average of $13, 15, 20$ and x is 18. What is the value of x?

A. 9

B. 15

C. 18

D. 24

22) What is the value of mode and median in the following set of numbers?

$$1, 2, 2, 5, 4, 4, 3, 3, 3, 1, 1$$

A. Mode: 1, 2 Median: 2

B. Mode: 1, 3 Median: 3

C. Mode: 2, 3 Median: 2

D. Mode: 1, 3 Median: 2.5

23) 5 less than twice a positive integer is 83. What is the integer?

A. 39

B. 41

C. 42

D. 44

24) If Joe was making $7.50 per hour and got a raise to $7.75 per hour, what percentage increase was the raise?

A. 1%

B. 2.67%

C. 3.33%

D. 6.66%

25) Which is the equivalent temperature of $104°F$ in Celsius? $(C = Celsius)$

$$C = \frac{5}{9}(F - 32)$$

A. 32
B. 38
C. 40
D. 44

Quantitative Comparisons

Direction: Questions 26 to 37 are Quantitative Comparisons Questions. Using the information provided in each question, compare the quantity in column A to the quantity in Column B. Choose on your answer sheet grid

- A if the quantity in Column A is greater
- B if the quantity in Column B is greater
- C if the two quantities are equal
- D if the relationship cannot be determined from the information given

26) A computer costs $270

Quantity A	Quantity B
A sales tax at 9% of the computer cost	$24.3

27)

Column A	Column B
$3^2 - 5^4$	$3^4 - 5^2$

28) $3x^4 - 45 = 723$

$$\frac{1}{5}y - \frac{4}{10} = \frac{3}{5}$$

Column A	Column B
x	y

29)

Column A	Column B
$\dfrac{\sqrt{81 - 49}}{\sqrt{16 - 9}}$	$\dfrac{(9 - 7)}{(4 - 3)}$

30)

Column A	Column B
$\sqrt{25} - \sqrt{9}$	$\sqrt{25 - 9}$

31) The sum of 3 consecutive integers is 54.

Column A	Column B
The largest of these integers	20

32)

Column A	Column B
The slope of the line $-4x + 4y = 1$	The slope of the line that passes through points $(3, 5)$ and $(1, 3)$

33) 10 percent of x is equal to 8 percent of y, where x and y are positive numbers.

Quantity A	Quantity B
x	y

34)

Quantity A	Quantity B
$(-4)^3$	4^3

35)

Quantity A	Quantity B
The least prime factor of 22	The least prime factor of 32

36)

Quantity A	Quantity B
$(1.25)^2(1.25)^6$	$(1.25)^8$

37) x is a positive number.

Quantity A	Quantity B
x^6	x^{11}

ISEE Middle Level Math
Practice Test 2

Section 2

47 questions

Total time for this section: 40 Minutes

You may NOT use a calculator for this test.

1) If $x =$ lowest common multiple of 30 and 35, then $\frac{x}{2} + 1$ equal to?
A. 210
B. 108
C. 106
D. 96

2) What is the value of x in the following equation?
$$3^x - 15 = 66$$
A. 3
B. 4
C. 5
D. 6

3) Which of the following is not synonym for 10^2?
A. 10 cubed
B. 10 squared
C. The square of 10
D. 10 to the second power

4) If angles A, B and C are angles of a parallelogram, what is the sum of the measures of the three angles?
A. $360\ degrees$
B. $180\ degrees$
C. $90\ degrees$
D. Cannot be determined

5) A swing moves from one extreme point (point A) to the opposite extreme point (point B) in 30 seconds. How long does it take that the swing moves 10 times from point A to point B and returns to point A?
A. $600\ seconds$
B. $300\ seconds$
C. $200\ seconds$
D. $100\ seconds$

6) There are 2 cars moving in the same direction on a road. A red car is $10\ km$ ahead of a blue car. If the speed of the red car is $50\ km\ per\ hour$ and the speed of the blue car is $1\frac{2}{5}$ of the red car, how many minutes will it take the blue car to catch the red car?
A. 8.5
B. 15
C. 30
D. 60

7) If the area of trapezoid is 100, what is the perimeter of the trapezoid?

A. 25
B. 28
C. 35
D. 45

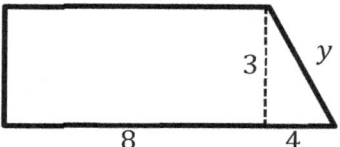

8) In two successive years, the population of a town is increased by 15% and 20%. What percent of the population is increased after two years?

A. 32%
B. 35%
C. 38%
D. 42%

9) In 1999, the average worker's income increased $2,000 per year starting from $24,000 annual salary. Which equation represents income greater than average?
(I = income, x = number of years after 1999)

A. $I > 2,000x + 24,000$
B. $I > -2,000x + 24,000$
C. $I < -2,000x + 24,000$
D. $I < 2,000x - 24,000$

10) Which of the following angles is obtuse?

A. $220\ degrees$
B. $340\ degrees$
C. $79\ degrees$
D. $110\ degrees$

11) $8 + 7 \times (-2) - [5 + 22 \times 4] \div 6 = ?$

A. 120
B. 88
C. -21.5
D. -20

12) What is ratio of perimeter of figure A to area of figure B?

A. $\frac{3}{8}$
B. $\frac{8}{3}$
C. $\frac{8}{5}$
D. $\frac{5}{8}$

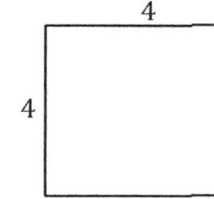

Fig. A

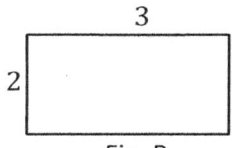

Fig. B

13) Mr. Jones saves $2,500 out of his monthly family income of $55,000. What fractional part of his income does he save?

A. $\frac{1}{22}$

B. $\frac{1}{11}$

C. $\frac{3}{25}$

D. $\frac{2}{15}$

14) Anita's trick–or–treat bag contains 12 pieces of chocolate, 18 suckers, 18 pieces of gum, 24 pieces of licorice. If she randomly pulls a piece of candy from her bag, what is the probability of her pulling out a piece of sucker

A. $\frac{1}{3}$

B. $\frac{1}{4}$

C. $\frac{1}{6}$

D. $\frac{1}{12}$

15) What is the difference in area between a 9 cm by 4 cm rectangle and a circle with diameter of 10 cm? ($\pi = 3$)

A. 49

B. 40

C. 39

D. 26

16) What is the value (values) of x in the following equation?
$$(x^2 + 4x + 4) = 100$$

A. $-8, 12$

B. $8, -12$

C. -8

D. 12

17) 120 is equal to?

A. $20 - (4 \times 10) + (6 \times 30)$

B. $\left(\frac{11}{8} \times 72\right) + (\frac{125}{5})$

C. $\left(\left(\frac{30}{4} + \frac{13}{2}\right) \times 7\right) - \frac{11}{2} + \frac{110}{4}$

D. $(2 \times 10) + (50 \times 1.5) + 15$

18) $\frac{15 \times 21}{8}$ is closest estimate to?

A. 39.4
B. 40.5
C. 42.6
D. 45.2

19) When a number is multiplied to itself and added by 10, the result is 35. What is the value of the number?

A. 5 and −5
B. 6 and −6
C. 5
D. 6

20) If Tom invests $1,000 at an annual rate of 4.5%, how much interest will he earn after two year?

A. $9,000
B. $4,500
C. $900
D. $90

21) How many possible outfit combinations come from six shirts, three slacks, and five ties?

A. 90
B. 60
C. 15
D. 14

22) What is the value of x in the equation $\frac{x}{8} + \frac{5}{8} = \frac{15}{4}$?

A. 25
B. 15
C. 8
D. 5

23) What is the absolute value of the quantity six minus nine?

A. −3
B. 15
C. −15
D. 3

24) If $y = 4ab + 3b^3$, what is y when $a = 2$ and $b = 3$?
A. 110
B. 105
C. 81
D. 36

25) Which of the following angles can represent the three angles of a triangle?
A. $20°, 80°, 90°$
B. $50°, 50°, 90°$
C. $10°, 110°, 60°$
D. $55°, 45°, 70°$

26) In the following equation, what is the value of $x - 2y$?
$$x + 3x - 10 = 2\left(\frac{3}{2}x + y\right) - 15$$
A. 5
B. -25
C. -5
D. 25

27) Two-kilogram banana and three-kilograms grapes cost $28. If the price of one-kilogram of banana is twice the price of one-kilogram of grape, how much does one kilogram banana cost?
A. $8
B. $4
C. $2
D. $1

28) Which is **NOT** a prime number?
A. 29
B. 43
C. 47
D. 121

29) How many tiles of $8 \ cm^2$ is needed to cover a floor of dimension $6 \ cm$ by $24 \ cm$?
A. 12
B. 18
C. 24
D. 36

30) There are three books, a red book, a blue book, and a white book. If the weight of the red book is $80\ g$ and it is 40% of the weight of the blue book, and the weight of the blue book is 125% of the weight of the white book, what is the weight of all three books?
A. $400\ g$
B. $300\ g$
C. $250\ g$
D. $200\ g$

31) Each of the x students in a team may invite up to 5 friends to a party. What is the maximum number of students and guests who might attend the party?
A. $5x + 5$
B. $5x$
C. $x + 5$
D. $6x$

32) Calculate the approximate circumference of the following circle. (diameter is 20)

A. $1,257$
B. 314
C. 63
D. 56

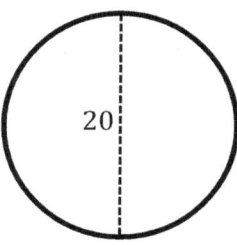

33) $270\ minutes = \cdots\ ?$
A. $5\ Hours$
B. $4.5\ Hours$
C. $3.5\ Hours$
D. $0.2\ Hours$

34) In the figure below, two lines are parallel. What is the value of angle x?

A. $35\ degree$
B. $92\ degree$
C. $120\ degree$
D. $145\ degree$

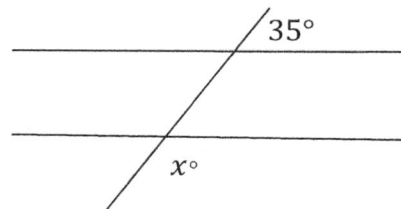

35) $\left(\left((-25) + 50\right) \times \frac{1}{5}\right) + (-10)?$
A. 5
B. 6
C. -5
D. -6

36) Adam drove 340 miles and it took him approximately 8 hours. How many miles per hour was his average speed?
A. about 44.5 miles per hour
B. about 42.5 miles per hour
C. about 36.5 miles per hour
D. about 31.5 miles per hour

37) Three people go to a restaurant. Their bill comes to $56.00. They decided to split the cost. One person pays $8.5, the next person pays 2 times that amount. How much will the third person have to pay?
A. $36.50
B. $30.50
C. $25.00
D. $20.00

38) What is 21,8210 in scientific notation?
A. 218.21×10^3
B. 21.821×10^4
C. 0.21821×10^6
D. 2.1821×10^5

39) What is the perimeter of the below right triangle?

A. 60
B. 30
C. 20
D. 10

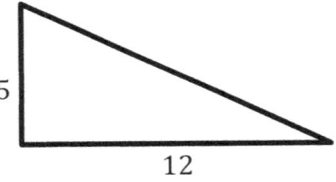

40) If 120% of a number is equal to 25% of 90, then what is the number?
A. 22.75
B. 18.75
C. 12.75
D. 10.75

41) What is the value of $(12 - 8)!$?
A. 20
B. 24
C. 27
D. 28

42) If x represents the greatest even multiple odd 11 less than 80, and y represents the least odd multiple of 9 greater than 30, what is $x + y$?
A. 111
B. 113
C. 122
D. 145

43) The average weight of 20 girls in a class is $50\ kg$ and the average weight of 30 boys in the same class is $60\ kg$. What is the average weight of all the 50 students in that class?
A. $56\ kg$
B. $58\ kg$
C. $59\ kg$
D. $59.5\ kg$

44) What is x in the following right triangle?

A. $\sqrt{388}$
B. 20
C. $\sqrt{401}$
D. 22

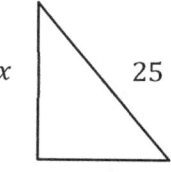

45) John traveled $150\ km$ in 6 hours and Alice traveled $180\ km$ in 4 hours. What is the ratio of the average speed of John to average speed of Alice?
A. $3 : 2$
B. $2 : 3$
C. $5 : 9$
D. $5 : 6$

46) An angle is equal to one fifth of its supplement. What is the measure of that angle?
A. 20
B. 30
C. 45
D. 60

47) What is the difference of smallest 4–digit number and biggest 4–digit number?

A. 6,666
B. 6,789
C. 8,888
D. 8,999

STOP

ISEE Middle Level Math Practice Tests Answers and Explanations

Now, it's time to review your results to see where you went wrong and what areas you need to improve!

ISEE Middle Level Math Practice Test 1 Answer Key											
Quantitative Reasoning						**Mathematics Achievement**					
1	A	17	C	33	A	1	A	17	C	33	A
2	B	18	A	34	B	2	A	18	A	34	D
3	B	19	A	35	C	3	D	19	B	35	B
4	C	20	C	36	C	4	D	20	A	36	B
5	B	21	A	37	D	5	B	21	D	37	D
6	A	22	D			6	B	22	A	38	D
7	B	23	D			7	D	23	D	39	B
8	A	24	D			8	D	24	D	40	D
9	D	25	B			9	C	25	C	41	A
10	D	26	A			10	A	26	C	42	A
11	C	27	B			11	D	27	C	43	B
12	A	28	A			12	A	28	B	44	D
13	B	29	C			13	B	29	D	45	A
14	A	30	C			14	D	30	D	46	D
15	B	31	C			15	D	31	C	47	A
16	B	32	A			16	B	32	B		

ISEE Middle Level Math Practice Test 2 Answer Key

Quantitative Reasoning

#		#		#	
1	B	17	C	33	B
2	B	18	B	34	B
3	B	19	C	35	B
4	D	20	A	36	C
5	A	21	D	37	A
6	B	22	B		
7	A	23	D		
8	A	24	C		
9	D	25	C		
10	C	26	C		
11	D	27	A		
12	A	28	B		
13	B	29	A		
14	A	30	B		
15	C	31	B		
16	A	32	C		

Mathematics Achievement

#		#		#	
1	C	17	C	33	B
2	B	18	A	34	D
3	A	19	A	35	C
4	D	20	D	36	B
5	A	21	A	37	B
6	C	22	A	38	D
7	B	23	D	39	B
8	C	24	B	40	B
9	A	25	C	41	B
10	D	26	C	42	A
11	C	27	A	43	A
12	B	28	D	44	B
13	A	29	B	45	C
14	B	30	A	46	B
15	C	31	D	47	D
16	B	32	C		

Score Your Test

ISEE scores are broken down by its four sections: Verbal Reasoning, Reading Comprehension, Quantitative Reasoning, and Mathematics Achievement. A sum of the three sections is also reported.

For the Middle Level ISEE, the score range is 760 to 940, the lowest possible score a student can earn is 760 and the highest score is 940 for each section. A student receives 1 point for every correct answer. There is no penalty for wrong or skipped questions.

The total scaled score for a Middle Level ISEE test is the sum of the scores for all sections. A student will also receive a percentile score of between 1-99% that compares that student's test scores with those of other test takers of same grade and gender from the past 3 years.

Use the next table to convert ISEE Middle level raw score to scaled score for application to 7th and 8th grade.

	ISEE Middle Level Scaled Scores									
Raw Score	Quantitative Reasoning		Mathematics Achievement		Raw Score	Quantitative Reasoning		Mathematics Achievement	
	7th Grade	8th Grade	7th Grade	8th Grade		7th Grade	8th Grade	7th Grade	8th Grade
0	760	760	760	760	26	900	885	885	865
1	770	765	770	765	27	905	890	885	865
2	780	770	780	770	28	910	895	890	870
3	790	775	790	775	29	910	900	890	870
4	800	780	800	780	30	915	905	895	875
5	810	785	810	785	31	920	910	895	875
6	820	790	820	790	32	925	915	900	880
7	825	795	825	795	33	930	920	900	880
8	830	800	830	800	34	930	925	905	885
9	835	805	835	805	35	935	930	905	885
10	840	810	840	810	36	935	935	910	890
11	845	815	845	815	37	940	940	910	890
12	850	820	850	820	38			915	895
13	855	825	855	825	39			920	900
14	860	830	855	830	40			925	905
15	865	835	860	835	41			925	910
16	870	840	860	840	42			930	915
17	875	845	865	840	43			930	920
18	880	845	865	845	44			935	925
19	880	850	870	845	45			935	930
20	885	855	870	850	46			940	935
21	885	860	875	850	47			940	940
22	890	865	875	855					
23	890	870	875	855					
24	895	875	880	860					
25	895	880	880	860					

ISEE Middle LEVEL Math Practice Test 1 Section 1

1) **Choice A is correct**

$$343 = 7^3 \rightarrow \frac{7^x}{7} = 7^3 \rightarrow 7^{x-1} = 7^3 \rightarrow x - 1 = 3 \rightarrow x = 4$$

2) **Choice B is correct**

In triangle sum of all angles equal to $180°$ then: $\quad y = 180° - (120° + 30°) =$ $180° - 150° = 30°$

3) **Choice B is correct**

$\frac{1}{3} \cong 0.33 \qquad \frac{4}{7} \cong 0.57 \qquad \frac{8}{12} \cong 0.66 \qquad \frac{3}{4} = 0.75$

4) **Choice C is correct**

One hour equal to 60 minutes then, $5 \ hours = 5 \times 60 = 300 \ minutes$

One minute equal to 60 seconds then, $300 \ minutes = 300 \times 60 = 18,000 \ seconds$

Distance that travel by object is: $\quad 0.4 \times 18,000 = 7,200 \ cm = 72 \ m$

5) **Choice B is correct**

Number of visiting fans: $\qquad \frac{2 \times 24,000}{5} = 9,600$

6) **Choice A is correct**

Circumference of circle$= 2\pi r = 2\pi \times \frac{17}{2} = 17\pi \sim 53.38 \ m$

7) **Choice B is correct**

The lowest common multiple of 24 and 36 is 72.

8) **Choice A is correct**

Area of circle with diameter 8 is: $\quad \pi r^2 = \pi \left(\frac{8}{2}\right)^2 = 16\pi,$ The area of shaded region is: $\frac{16\pi}{4} = 4\pi$

9) **Choice D is correct**

30% of $200 = \frac{30}{100} \times 200 = 60,$ Final sale price is: $\quad 200 - 60 = \$140$

10) **Choice D is correct**

To find the discount, multiply the number by $(100\% - rate \ of \ discount)$.

Therefore, for the first discount we get: $(300) \ (100\% - 15\%) = (300) \ (0.85)$

For the next 25% discount: $(300)\,(0.85)\,(0.75)$

11) Choice C is correct

The amount of petrol consumed after x hours is: $\;5x$, Petrol remaining: $70 - 5x$

12) Choice A is correct

$4 + x + 8\left(\dfrac{x}{4}\right) = 2x + 12 \rightarrow 4 + x + 2x = 2x + 12 \rightarrow x = 8$

13) Choice B is correct

The area of trapezoid is: $\;\left(\dfrac{16+1}{2}\right)x = 136 \rightarrow 17x = 136 \rightarrow x = 8$

14) Choice A is correct

$\dfrac{3}{5}$ of $280 = \dfrac{3}{5} \times 280 = 168$, $\dfrac{1}{2}$ of $168 = \dfrac{1}{2} \times 168 = 84$, $\dfrac{1}{3}$ of $84 = \dfrac{1}{3} \times 84 = 28$

15) Choice B is correct

$6 + 3x \le 21 \rightarrow 3x \le 21 - 6 \rightarrow 3x \le 15 \rightarrow x \le \dfrac{15}{3} \rightarrow x \le 5, \quad$ Then: $a = 5$

16) Choice B is correct

$One\ liter = 1{,}000\ cm^3 \rightarrow 7\ liters = 7{,}000\ cm^3$, $7{,}000 = 25 \times 5 \times h \rightarrow h = \dfrac{7{,}000}{125} = 56\ cm$

17) Choice C is correct

$4f + 4g = 4x - 2y \rightarrow 4f + 4(2y - 6x) = 4x - 2y \rightarrow 4f + 8y - 24x = 4x - 2y \rightarrow$

$4f = 28x - 10y \rightarrow 2f = 14x - 5y$

18) Choice A is correct

$\dfrac{-\frac{13}{3} \times \frac{4}{5}}{\frac{10}{30}} = -\dfrac{\frac{13 \times 4}{3 \times 5}}{\frac{10}{30}} = -\dfrac{\frac{52}{15}}{\frac{10}{30}} = -\dfrac{52 \times 30}{15 \times 10} = -10.4$

19) Choice A is correct

Use Pythagorean theorem to find the value of c: $\quad a^2 + b^2 = c^2 \rightarrow 5^2 + 12^2 = c^2 \rightarrow$

$169 = c^2 \rightarrow c = 13$. Perimeter of parallelogram$= (9 + 5 + 13) \times 2 = 54$

20) Choice C is correct

$\dfrac{6}{50} \times 100 = \dfrac{6}{5} \times 10 = 12\%$

21) Choice A is correct

Let x be the integer. Then: $2x - 3 = 71$, Add 3 both sides: $2x = 74$, Divide both sides by 2: $x = 37$

22) Choice D is correct

$\dfrac{0.35}{8.5} \times 100 = 4.11 \approx 4.00$

23) Choice D is correct

Plug in 140 for F and then solve for C. $C = \frac{5}{9}(F - 32) \Rightarrow C = \frac{5}{9}(140 - 32) \Rightarrow$

$C = \frac{5}{9}(108) = 60$

24) Choice D is correct

$Average = \frac{sum\ of\ terms}{number\ of\ terms} \Rightarrow 20 = \frac{14+16+21+x}{4} \Rightarrow 80 = 51 + x \Rightarrow x = 29$

25) Choice B is correct

First write the numbers in the order: 2,2,2,3,3,4,4,5,5,6,6

The mode of numbers is: 2   median is: 4

26) Choice A is correct.

Column A: Simplify. $\frac{\sqrt{64-48}}{\sqrt{25-9}} = \frac{\sqrt{16}}{\sqrt{16}} = 1$

Column B: $\frac{(7-4)}{(8-3)} = \frac{3}{5}$

27) Choice B is correct

$2x^5 - 9 = 477 \rightarrow 2x^5 = 477 + 9 = 486 \rightarrow x^5 = \frac{486}{2} = 243 \rightarrow x = \sqrt[5]{243} = \sqrt[5]{3^5} = 3$

$\frac{1}{3} - \frac{y}{5} = -\frac{7}{15} \rightarrow \frac{y}{5} = \frac{1}{3} + \frac{7}{15} = \frac{5+7}{15} = \frac{12}{15} = \frac{4}{5} \rightarrow y = 5 \times \frac{4}{5} = 4$

28) Choice A is correct.

Column A: First, find the integers. Let x be the smallest integer. Then the integers are x, $(x + 1)$, and $(x + 2)$. The sum of the integers is -45. Then:

$x + x + 1 + x + 2 = -45 \rightarrow 3x + 3 = -45 \rightarrow 3x = -48 \rightarrow x = -16$

The smallest integer is -16, therefore, the largest integer is bigger than that.

29) Choice C is correct.

Column A: $4^2 - 2^4 = 16 - 16 = 0$

Column B: $2^4 - 4^2 = 16 - 16 = 0$

30) Choice C is correct.

Column A: 8% of the computer cost is 20: $8\% \times 250 = 0.08 \times 250 = 20$

Column B: 20

31) Choice C is correct.

Column A: The slope of the line $4x + 2y = 7$ is -2.

Write the equation in slope intercept form. $4x + 2y = 7 \rightarrow 2y = -4x + 7 \rightarrow y = -2x + \frac{7}{2}$

Column B: The slope of the line that passes through points $(2, 5)$ and $(3, 3)$:

Use slope formula: $slope\ of\ a\ line = \frac{y_2 - y_1}{x_2 - x_1} = \frac{3-5}{3-2} = -2$

32) Choice A is correct

prime factoring of 55 is: 5×11 , prime factoring of 210 is: $2 \times 3 \times 5 \times 7$

Quantity $A = 5$ and Quantity $B = 2$

33) Choice A is correct.

Column A: Simplify. $\sqrt{144 - 81} = \sqrt{63}$

Column B: $\sqrt{144} - \sqrt{81} = 12 - 9 = 3$

$\sqrt{63}$ is bigger than 3. ($\sqrt{9} = 3$)

34) Choice B is correct

6% of x = 5% of $y \rightarrow 0.06\,x = 0.05\,y \rightarrow x = \frac{0.05}{0.06}y \rightarrow x = \frac{5}{6}y$, therefore, y is bigger than x.

35) Choice C is correct

Simplify both quantities.

Quantity A: $\quad (-5)^4 = (-5) \times (-5) \times (-5) \times (-5) = 625$

Quantity B: $5 \times 5 \times 5 \times 5 = 625$
The two quantities are equal.

36) Choice C is correct.

Use exponent "product rule": $x^n \times x^m = x^{n+m}$

Quantity A: $\quad (1.88)^4(1.88)^8 = (1.88)^{4+8} = (1.88)^{12}$

Quantity B: $(1.88)^{12}$
The two quantities are equal.

37) Choice D is correct.

Choose different values for x and find the value of quantity A and quantity B.

$x = 1$, then: Quantity A: $\ x^{10} = 1^{10} = 1$

Quantity B: $\ x^{20} = 1^{20} = 1$

The two quantities are equal.

$x = 2$, then: Quantity A: $\ x^{10} = 2^{10}$

Quantity B: $\ x^{20} = 2^{20}$

Quantity B is greater.

Therefore, the relationship cannot be determined from the information given

ISEE Middle LEVEL Math Practice Test 1 Section 2

1) Choice A is correct

20 cubed is: $20^3 = 8,000$

2) Choice A is correct

$3^x + 28 = 55 \rightarrow 3^x = 55 - 28 = 27$ and $27 = 3^3, 3^x = 3^3 \rightarrow x = 3$

3) Choice D is correct

All angles in a parallelogram sum up to 360 degrees. Since, we only have 2 angles, therefore the answer cannot be determined.

4) Choice D is correct

Prime factorizing of $10 = 2 \times 5$,   Prime factorizing of $35 = 5 \times 7$

$x = LCM = 2 \times 5 \times 7 = 70$, $\frac{70}{5} + 2 = 14 + 2 = 16$

5) Choice B is correct

The area of trapezoid is: $\left(\frac{8+12}{2}\right) \times x = 30 \rightarrow 10x = 30 \rightarrow x = 3$

$y = \sqrt{3^2 + 4^2} = 5$,   Perimeter is: $12 + 3 + 8 + 5 = 28$

6) Choice B is correct

Swing moves once from point A to point B and returns to point A is: $20 + 20 = 40$ seconds

Therefore, for ten times: $5 \times 40 = 200$ seconds

7) Choice D is correct

Speed of the blue car: $1\frac{2}{5} \times 40 = 56$, Difference of the cars' speed: $56 - 40 = 16$, The red car is $12 \; km$ ahead of a blue car. Therefore, it takes the blue car 45 minutes to catch the red car. $\frac{12}{16} = \frac{3}{3}$ Hour $= 45$ minutes

8) Choice D is correct

The population is increased by 10% and 25%. 10% increase changes the population to 110% of original population. For the second increase, multiply the result by 125%.

$(1.10) \times (1.25) = 1.37 = 137\%$,   37 percent of the population is increased after two years.

9) Choice C is correct

Use PEMDAS (order of operation): $5 + 8 \times (-2) - [4 + 22 \times 5] \div 6 = 5 + 8 \times (-2) - [4 + 110] \div 6 = 5 + 8 \times (-2) - [114] \div 6 = 5 + (-16) - 19 = 5 + (-16) - 19 = -11 - 19 = -30$

10) Choice A is correct

Let x be the number of years. Therefore, \$2,500 per year equals $2,500x$. starting from \$26,000 annual salary means you should add that amount to $2,500x$. Income more than that is: $I > 2,500\, x + 26,000$

11) Choice D is correct

Angle between 90° and 180° is called obtuse angle.

12) Choice A is correct

2,500 out of 65,000 equals to $\frac{2,500}{65,000} = \frac{25}{650} = \frac{1}{26}$

13) Choice B is correct

$Probability = \frac{number\ of\ desired\ outcomes}{number\ of\ total\ outcomes} = \frac{19}{13+19+19+25} = \frac{19}{76} = \frac{1}{4}$

14) Choice D is correct

Only choice D equals 110: $(2 \times 10) + (50 \times 1.5) + 15 = 20 + 75 + 15 = 110$

15) Choice D is correct

The area of rectangle is: $8 \times 4 = 32\ cm^2$, The area of circle is: $\pi r^2 = \pi \times \left(\frac{8}{2}\right)^2 = 3 \times 16 = 48$, Difference in area is: $48 - 32 = 16$

16) Choice B is correct

$x^2 + 2x + 1 = (x+1)^2 \to (x+1)^2 = 64 \to x + 1 = 8 \to x = 7\ or\ x + 1 = -8 \to x = -9$

17) Choice C is correct

Perimeter $A = 4 \times 6 = 24$, $\qquad$ Area $B = 2 \times 4 = 8$, $\frac{24}{8} = 3$

18) Choice A is correct

$\frac{18 \times 21}{5} = \frac{378}{5} = 75.6$

19) Choice B is correct

To find the number of possible outfit combinations, multiply number of options for each factor: $4 \times 2 \times 5 = 40$

20) Choice A is correct

Let x be the number, then: $x^2 + 9 = 25 \to x^2 = 16 \to x^2 - 16 = 0 \to (x+4)(x-4) = 0 \to x = 4\ or\ x = -4$

21) Choice D is correct

8% of $2,000 = \frac{8}{100} \times 2,000 = \$160$

22) Choice A is correct

$\frac{x}{4} + \frac{5}{4} = \frac{30}{6} \to \frac{x}{4} = 5 - \frac{5}{4} = \frac{20-5}{4} = \frac{15}{4} \to \frac{x}{4} \to x = 4 \times \frac{15}{4} = 15$

23) Choice D is correct

$y = 5ab + 3b^3$, Plug in the values of a and b in the equation: $a = 3$ and $b = 2$

$y = 5\ (3)(2) + 3\ (2)^3 = 30 + 3(8) = 30 + 24 = 54$

24) Choice D is correct

$|6 - 10| = |-4| = 4$

25) Choice C is correct

The angles of an equilateral triangle are $60, 60, 60$ degrees.

26) Choice C is correct

$9x - 10 = 5\left(\dfrac{4}{5}x - y\right) + 5 \rightarrow 9x - 10 = 4x - 5y + 5 \rightarrow 5x - 10 = -5y + 5$

$\rightarrow 5x = -5y + 15 \rightarrow x = -y + 3 \rightarrow x + y = 3$

27) Choice C is correct

The area of the floor is: $7cm \times 27\ cm = 189\ cm^2$, The number is tiles needed $= 189 \div 3 = 63$

28) Choice B is correct

Let x be price of one-kilogram of apple and y be price of one-kilogram of orange, then:

$x = 2y, 2x + 3y = 21 \rightarrow 2(2y) + 3y = 21 \rightarrow 7y = 21 \rightarrow y = \dfrac{21}{7} = 3 \rightarrow x = 2 \times 3 = 6$

29) Choice D is correct

122 is not prime number, it is divided by 2

30) Choice D is correct

Since, each of the x students in a team may invite up to 6 friends, the maximum number of people in the party is 7 times x or $7x$. ($one\ student + 6\ friends = 7\ people$)

31) Choice C is correct

Perimeter$= 2\pi r = 2 \times \pi \times \dfrac{10}{2} = 10 \approx 31.4 \approx 31$

32) Choice B is correct

$60\ minutes = 1\ Hour \rightarrow \dfrac{280}{60} = 4.6\ Hours$

33) Choice A is correct

Let x be the weight of the red box. Then: Weight of blue box: $0.8x = 60 \rightarrow x = \dfrac{60}{0.8} = 75$,

Weight of yellow box $1.20y = 75 \rightarrow y = 62.5$

The weight of all boxes: $60 + 75 + 62.5 = 197.5$

34) Choice D is correct

$180° - 40° = 140°$

35) Choice B is correct

Average speed: $\dfrac{350}{9} = 38.8\ miles\ per\ hour$

36) Choice B is correct

Let x be the price that third person has to pay then; $58 = 7.5 + (2 \times 7.5) + x \rightarrow$

$x = 58 - 22.5 = 35.5$

37) Choice D is correct

$$\left(\left((-15) + 40\right) \times \frac{1}{5}\right) + (-15) = \left((25) \times \frac{1}{5}\right) - 15 = 5 - 15 = -10$$

38) Choice D is correct

$138,210 = 1.3821 \times 10^5$

39) Choice B is correct

$(10 - 6)! = 4! = 4 \times 3 \times 2 \times 1 = 24$

40) Choice D is correct

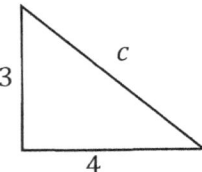

$c = \sqrt{4^2 + 3^2} = \sqrt{25} = 5$, Perimeter is: $\quad 5 + 3 + 4 = 12$

41) Choice A is correct

Let x be the number then, 150% of $x = 1.5x$, $1.5x = 0.30 \times 80 = 24 \rightarrow x = \frac{24}{1.5} = 16$

42) Choice A is correct

Number of employees with Diploma: $\qquad \frac{4 \times 24}{1} = 96$

That means there are 24 employees with Bachelor's Degree and 96 employees with high school diploma. To get the ratio of employees with Bachelor's Degree to the number of employees with High School Diploma 3 to 4, there should be 32 employees with high school diploma. So, 40 ($96 - 32 = 64$) employees with High School Diploma should be moved to other departments.

43) Choice B is correct

Use Pythagorean theorem: $x^2 + 20^2 = 25^2 \rightarrow x^2 = 25^2 - 20^2 \rightarrow x = \sqrt{25^2 - 20^2} \rightarrow x = \sqrt{625 - 400} = \sqrt{225} = 15$

44) Choice D is correct

$average = \frac{sum\ of\ terms}{number\ of\ terms}$, The sum of the weight of all girls is: $20 \times 65 = 1,300\ kg$

The sum of the weight of all boys is: $35 \times 72 = 2,520\ kg$, The sum of the weight of all students is: $1,300 + 2,520 = 3,820\ kg$, $Average = \frac{3,820}{55} = 69.45$

45) Choice A is correct

The sum of supplement angles is 180. Let x be that angle. Therefore, $x + 9x = 180$, $10x = 180$, divide both sides by 10: $x = 18$

46) Choice D is correct

Smallest 5–digit number is 10,000, and biggest 5–digit number is 99,999. The difference is: 89,999 ,

47) Choice A is correct

The average speed of john is: $140 \div 5 = 28$, The average speed of Alice is: $210 \div 3 = 70$

Write the ratio and simplify. $28 : 70 \Rightarrow 2 : 5$

ISEE Middle Level Math Practice Test 2 Section 1

1) Choice B is correct

Number of visiting fans: $\frac{2 \times 25,000}{5} = 10,000$

2) Choice B is correct

In triangles sum of all angles equal to $180°$, then: $y = 180° - (100° + 28.5°) =$

$100° - 128.5° = 51.5°$

3) Choice B is correct

$\frac{2}{3} \cong 0.67$ $\frac{5}{7} \cong 0.71$ $\frac{8}{11} \cong 0.72$ $\frac{3}{4} = 0.75$

4) Choice D is correct

One hour equal to 60 minutes then, 4 hours= $4 \times 60 = 240$ minutes

One minute equal to 60 seconds then, 240 minutes= $240 \times 60 = 14,400$ seconds

Distance that travel by object is: $0.3 \times 14,400 = 4,320 \, cm = 43.2 \, m$

5) Choice A is correct

$243 = 3^5 \rightarrow \dfrac{3^x}{9} = 3^5 \rightarrow \dfrac{3^x}{3^2} = 3^5 \rightarrow 3^{x-2} = 3^5 \rightarrow x - 2 = 5 \rightarrow x = 7$

6) Choice B is correct

40% off is: 40% of $40 = \dfrac{40}{100} \times 40 = \$16$, then, the sales price of the sweater is $24.

5% of $24 = \dfrac{5}{100} \times 24 = \$1.20$. Ava pays: $\$24 + \$1.2 = \$25.20$

7) Choice A is correct

20% of $300 = \dfrac{20}{100} \times 300 = 60$,   Final sale price is: $300 - 60 = \$240$

8) Choice A is correct

Circumference of circle= $2\pi r = 2\pi \times \dfrac{15}{2} = 15\pi = 15 \times 3.14 = 47.1 \, m$

9) Choice D is correct

Area of circle with diameter 12 is: $\pi r^2 = \pi \left(\frac{12}{2}\right)^2 = 36\pi$, the area of shaded region is:

$\frac{36}{4} = 9\pi$

10) Choice C is correct

The amount of petrol consumed after x hours is: $6x$, Petrol remaining: $80 - 6x$

11) Choice D is correct

To find the discount, multiply the number by $(100\% - rate\ of\ discount)$.

Therefore, for the first discount we get: $(200)(100\% - 15\%) = (200)(0.85)$

For the next 15% discount: $(200)(0.85)(0.85)$

12) Choice A is correct

$\frac{3}{4}$ of $290 = \frac{3}{4} \times 300 = 225$, $\frac{2}{5}$ of $225 = \frac{2}{5} \times 225 = 90$, $\frac{1}{3}$ of $90 = \frac{1}{3} \times 90 = 30$

13) Choice B is correct

$7 + 2x \leq 15 \rightarrow 2x \leq 15 - 7 \rightarrow 2x \leq 8 \rightarrow x \leq \frac{8}{2} \rightarrow x \leq 4$, Then: $a = 4$

14) Choice A is correct

The area of trapezoid is: $\left(\frac{15+18}{2}\right) x = 132 \rightarrow 16.5x = 132 \rightarrow x = 8$

15) Choice C is correct

$3 + x + 6\left(\frac{x}{2}\right) = 2x + 10 \rightarrow 3 + x + 3x = 2x + 10 \rightarrow 2x = 7 \rightarrow x = \frac{7}{2}$

16) Choice A is correct

$One\ liter = 1,000\ cm^3 \rightarrow 6\ liters = 6,000\ cm^3$, $6,000 = 15 \times 5 \times h \rightarrow h = \frac{6,000}{75} = 80\ cm$

17) Choice C is correct

$3f + 2g = 3x + y \rightarrow 3f + 2(2y - 3x) = 3x + y \rightarrow 3f + 4y - 6x = 3x + y \rightarrow$

$3f = 9x - 3y \rightarrow f = 3x - y$

18) Choice B is correct

$c = \sqrt{4^2 + 3^2} = \sqrt{25} = 5$

Perimeter of parallelogram$= (9 + 3 + 5) \times 2 = 34$

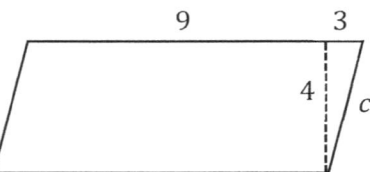

ISEE Middle Level Math Workbook 2020 - 2021

19) Choice C is correct

$$\frac{8}{40} \times 100 = \frac{8}{4} \times 10 = 20\%$$

20) Choice A is correct

$$\frac{-\frac{11}{2} \times \frac{3}{5}}{\frac{11}{30}} = -\frac{\frac{11 \times 3}{2 \times 5}}{\frac{11}{30}} = -\frac{\frac{33}{10}}{\frac{11}{30}} = -\frac{33 \times 30}{11 \times 10} = -9$$

21) Choice D is correct

$$Average = \frac{sum\ of\ terms}{number\ of\ terms} \Rightarrow 18 = \frac{13 + 15 + 20 + x}{4} \Rightarrow 72 = 48 + x \Rightarrow x = 24$$

22) Choice B is correct

Write the numbers in order: 1, 1, 1, 2, 2, 3, 3, 3, 4, 4, 5

The mode of numbers is: 1 and 3   median is: 3

23) Choice D is correct

Let x be the integer. Then: $2x - 5 = 83$, Add 5 both sides: $2x = 88$, Divide both sides by 2:

$$x = 44$$

24) Choice C is correct

$$\frac{0.25}{7.5} \times 100 = 3.33$$

25) Choice C is correct

Plug in 104 for F and then solve for C. $C = \frac{5}{9}(F - 32) \Rightarrow C = \frac{5}{9}(104 - 32) \Rightarrow$

$$C = \frac{5}{9}(72) = 40$$

26) Choice C is correct

Column A: 9% of the computer cost is 24.3: $9\% \times 270 = 0.09 \times 270 = 24.3$

Column B: 24.3

27) Choice B is correct.

Column A: $3^2 - 5^4 = 9 - 625 = -616$, Column B: $3^4 - 5^2 = 81 - 25 = 56$

28) Choice B is correct.

$$3x^4 - 45 = 723 \rightarrow 3x^4 = 723 + 45 = 768 \rightarrow x^4 = \frac{768}{3} = 256 \rightarrow x = \sqrt[4]{256} = \sqrt[4]{4^4} = 4$$

$$\frac{1}{5}y - \frac{4}{10} = \frac{3}{5} \rightarrow \frac{y}{5} = \frac{3}{5} + \frac{4}{10} = \frac{6+3}{10} = \frac{10}{10} = 1 \rightarrow y = 5 \times 1 = 5$$

29) Choice A is correct.

Column A: Simplify. $\frac{\sqrt{81-49}}{\sqrt{16-9}} = \frac{\sqrt{32}}{\sqrt{7}} = \sqrt{\frac{32}{7}}$ which is bigger than 2 ($\sqrt{4} = 2$ and $\frac{32}{7}$ is bigger than

4).   Column B: $\frac{(9-7)}{(4-3)} = \frac{2}{1} = 2$

30) Choice B is correct.

Column B: Simplify. $\sqrt{25-9} = \sqrt{16} = 4$

Column A: $\sqrt{25} - \sqrt{9} = 5 - 3 = 2$, $\sqrt{14}$ is bigger than 2. ($\sqrt{4} = 2$)

31) Choice B is correct.

Column A: First, find the integers. Let x be the smallest integer. Then the integers are x, $(x + 1)$, and $(x + 2)$. The sum of the integers is 54. Then:

$x + x + 1 + x + 2 = 54 \rightarrow 3x + 3 = 54 \rightarrow 3x = 51 \rightarrow x = 17$. The smallest integer is 17, therefore, the largest integer is 19 which is less than 20.

32) Choice C is correct.

Column A: Write the equation in slope intercept form. $-4x + 4y = 1 \rightarrow 4y = 4x + 1 \rightarrow$

$$y = x + \frac{1}{4}$$

Column B: The slope of the line that passes through points $(1, 3)$ and $(3, 5)$:

Use slope formula: $slope\ of\ a\ line = \frac{y_2 - y_1}{x_2 - x_1} = \frac{5-3}{3-1} = 1$

33) Choice B is correct

$10\%\ of\ x = 8\%$ of $y \rightarrow 0.1\ x = 0.08\ y \rightarrow x = \frac{0.08}{0.1}y \rightarrow x = \frac{8}{10}y$, therefore, y is bigger than x.

34) Choice B is correct

Quantity A: $(-4)^3 = (-4) \times (-4) \times (-4) = -64$,   Quantity B: $4 \times 4 \times 4 = 64$

Quantity B is greater.

35) Choice C is correct

Prime factoring of 22 is: 2×11, prime factoring of 32 is: $2 \times 2 \times 2 \times 2 \times 2$

Quantity A = 2 and Quantity B = 2

36) Choice C is correct.

Use exponent "product rule": $x^n \times x^m = x^{n+m}$, Quantity A: $(1.25)^2(1.25)^6 = (1.25)^{2+6} = (1.25)^8$, Quantity B: $(1.25)^8$, The two quantities are equal.

37) Choice D is correct.

Plug in different values for x and check quantity A and B. Let's choose 1 for x. Then:

Quantity A: $x^6 = 1^6 = 1$　　Quantity B: $x^{11} = 1^{11} = 1$

Now, let's choose 2 for x. Then: $2^6 < 2^{11}$

The relationship cannot be determined from the information given

ISEE Middle Level Math Practice Test 2 Section 2

1)　Choice C is correct

Prime factorizing of $30 = 2 \times 3 \times 5$,　　　Prime factorizing of $35 = 5 \times 7$

$x = LCM = 2 \times 3 \times 5 \times 7 = 210, \frac{210}{2} + 1 = 105 + 1 = 106$

2)　Choice B is correct

$3^x - 15 = 66 \rightarrow 3^x = 66 + 15 = 81$　　　and　　$81 = 3^4, 3^x = 3^4 \rightarrow x = 4$

3)　Choice A is correct

10 cubed is:　$10^3 = 1,000$

4)　Choice D is correct
All angles in a parallelogram sum up to 360 degrees. Since, we only have 3 angles, therefore the answer cannot be determined.

5)　Choice A is correct

Swing moves once from point A to point B and returns to point A is: $30 + 30 = 60 \; seconds$

Therefore, for ten times:　　$10 \times 60 = 600$ seconds

6)　Choice C is correct

$1\frac{2}{5} = \frac{7}{5} = 1.4$, Speed of the blue car: $1.4 \times 50 = 70$, Difference of the cars' speed:

$70 - 50 = 20$, The red car is $10 \; km$ ahead of a blue car. Therefore, it takes the blue car 30 minutes to catch the red car. $\frac{10}{20} = 0.5$ Hour $= 30$ minutes

7)　Choice B is correct

$y = \sqrt{3^2 + 4^2} = 5$,　　　　Perimeter is:　$12 + 3 + 8 + 5 = 28$

8) Choice C is correct

The population is increased by 15% and 20%. 15% increase changes the population to 115% of original population. For the second increase, multiply the result by 120%.

$(1.15) \times (1.20) = 1.38 = 138\%$,   38 percent of the population is increased after two years.

9) Choice A is correct

Let x be the number of years. Therefore, \$2,000 per year equals $2,000x$. starting from \$24,000 annual salary means you should add that amount to $2,000x$. Income more than that is: $I > 2,000\,x + 24,000$

10) Choice D is correct

Angle between $90°$ and $180°$ is called obtuse angle.

11) Choice C is correct

Use PEMDAS (order of operation):

$$8 + 7 \times (-2) - [5 + 22 \times 4] \div 6 = 8 + 7 \times (-2) - [5 + 88] \div 6$$
$$= 8 + 7 \times (-2) - [93] \div 6 = 8 + (-14) - 15.5 = 8 + (-14) - 15.5$$
$$= -6 - 15.5 = -21.5$$

12) Choice B is correct

Perimeter $A = 4 \times 4 = 16$,   Area $B = 2 \times 3 = 6$, $\frac{16}{6} = \frac{8}{3}$

13) Choice A is correct

2,500 out of 55,000 equals to $\frac{2,500}{55,000} = \frac{25}{550} = \frac{1}{22}$

14) Choice B is correct

$Probability = \frac{number\ of\ desired\ outcomes}{number\ of\ total\ outcomes} = \frac{18}{12+18+18+2} = \frac{18}{72} = \frac{1}{4}$

15) Choice C is correct

The area of rectangle is: $9 \times 4 = 36\ cm^2$, The area of circle is: $\pi r^2 = \pi \times (\frac{10}{2})^2 = 3 \times 25 = 75$, Difference in area is: $75 - 36 = 39$

16) Choice B is correct
$x^2 + 4x + 4 = (x + 2)^2 \rightarrow (x + 2)^2 = 100 \rightarrow x + 2 = 10 \rightarrow x = 8\ or\ x + 2 = -10 \rightarrow x = -12$

17) Choice C is correct
Only choice C is equal to 120.
$$\left(\left(\frac{30}{4} + \frac{13}{2}\right) \times 7\right) - \frac{11}{2} + \frac{110}{4} = \left(\left(\frac{30 + 26}{4}\right) \times 7\right) - \frac{11}{2} + \frac{55}{2} = \left(\left(\frac{56}{4}\right) \times 7\right) + \frac{55 - 11}{2}$$
$$= (14 \times 7) + \frac{44}{2} = 98 + 22 = 120$$

ISEE Middle Level Math Workbook 2020 - 2021

18) Choice A is correct

$$\frac{15 \times 21}{8} = \frac{315}{8} = 39.375 \sim 39.4$$

19) Choice A is correct

Let x be the number, then:

$$x^2 + 10 = 35 \rightarrow x^2 = 25 \rightarrow x^2 - 25 = 0 \rightarrow (x+5)(x-5) = 0 \rightarrow x = 5 \ or \ x = -5$$

20) Choice D is correct

4.5% of $\$1,000 = \frac{4.5}{100} \times 1,000 = \$45$. For two years the interest in $\$90$.

21) Choice A is correct

To find the number of possible outfit combinations, multiply number of options for each factor: $6 \times 3 \times 5 = 90$

22) Choice A is correct

$$\frac{x}{8} + \frac{5}{8} = \frac{15}{4} \rightarrow \frac{x}{8} = \frac{15}{4} - \frac{5}{8} = \frac{30-5}{8} = \frac{25}{8} = \rightarrow x = 25$$

23) Choice D is correct

$$|6 - 9| = |-3| = 3$$

24) Choice B is correct

$y = 4ab + 3b^3$, Plug in the values of a and b in the equation: $a = 2$ and $b = 3$

$$y = 4\,(2)\,(3)\ +\ 3\,(3)^3 = 24\ +\ 3(27) = 24 + 81 = 105$$

25) Choice C is correct

All three angles in a triangle add up to 180 degrees. Only choice C represents three angles of a triangle.

26) Choice C is correct

$$x + 3x - 10 = \left(2 \times \left(\frac{3}{2}x + y\right)\right) - 15 \rightarrow 4x - 10 = \left(2 \times \frac{3}{2}x\right) + 2y - 15 \rightarrow 4x - 10$$
$$= 3x + 2y - 15 \rightarrow 4x - 3x - 2y = 10 - 15 \rightarrow x - 2y = -5$$

27) Choice A is correct

Let x be price of one-kilogram of banana and y be price of one-kilogram of grape, then: $x = 2y$

$$2x + 3y = 28 \rightarrow 2(2y) + 3y = 28 \rightarrow 7y = 28 \rightarrow y = \frac{28}{7} = 4 \rightarrow x = 2 \times 4 = 8$$

28) Choice D is correct

121 is not prime number, it is divided by 11.

29) Choice B is correct

The area of the floor is: $6\ cm \times 24\ cm = 144\ cm^2$, The number is tiles needed $= 144 \div 8 = 18$

30) Choice A is correct

Weight of the blue book$= \frac{80}{0.4} = 200\ g$,   Weight of the white book$= \frac{200}{1.25} = 160\ g$

Weight of all three books: $80 + 200 + 160 = 400g$

31) Choice D is correct

Since, each of the x students in a team may invite up to 5 friends, the maximum number of people in the party is 6 times x or $6x$. $(one\ student + 5\ friends\ =\ 6\ people)$

32) Choice C is correct

Perimeter$= 2\pi r = 2 \times \pi \times \frac{20}{2} = 20 \approx 62.8 \approx 63$

33) Choice B is correct

$60\ minutes\ = 1 Hours \rightarrow \dfrac{270}{60} = 4.5\ hours$

34) Choice D is correct

$180° - 35° = 145°$

35) Choice C is correct

$$\left(\left((-25) + 50\right) \times \frac{1}{5}\right) + (-10) = \left((25) \times \frac{1}{5}\right) - 10 = 5 - 10 = -5$$

36) Choice B is correct

Average speed: $\dfrac{340}{8} = 42.5$ miles per hour

37) Choice B is correct

Let x be the price that third person has to pay then; $56 = 8.5 + (2 \times 8.5) + x \rightarrow$

$x = 56 - 25.5 = 30.5$

38) Choice D is correct

$218,210 = 2.1821 \times 10^5$

39) Choice B is correct

$c = \sqrt{5^2 + 12^2} = \sqrt{169} = 13$, Perimeter is: $5 + 12 + 13 = 30$

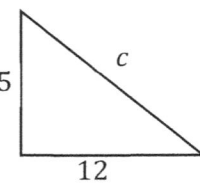

40) Choice B is correct

Let x be the number then, 120% of $x = 1.2x$, $1.2x = 0.25 \times 90 = 22.5 \rightarrow x = \frac{22.5}{1.2} = 18.75$

41) Choice B is correct

$(12 - 8)! = 4! = 4 \times 3 \times 2 \times 1 = 24$

42) Choice A is correct

The greatest even multiple odd 11 less than 80 is 66 (number 77 is odd) and the least odd multiple of 9 greater than 30 is 45. Then: $x + y = 66 + 45 = 111$

43) Choice A is correct

$average = \frac{sum\ of\ terms}{number\ of\ terms}$, The sum of the weight of all girls is: $20 \times 50 = 1,000\ kg$

The sum of the weight of all boys is: $30 \times 60 = 1,800\ kg$, The sum of the weight of all students is: $1,000 + 1,800 = 2,800\ kg$, $Average = \frac{2,800}{50} = 56$

44) Choice B is correct

Use Pythagorean Theorem: $x^2 + 15^2 = 25^2 \rightarrow x^2 = 25^2 - 15^2 \rightarrow x = \sqrt{25^2 - 15^2} \rightarrow$

$$x = \sqrt{625 - 225} = \sqrt{400} = 20$$

45) Choice C is correct

The average speed of john is: $150 \div 6 = 25$, The average speed of Alice is: $180 \div 4 = 45$

Write the ratio and simplify. $25:45 \Rightarrow 5:9$

46) Choice B is correct

The sum of supplement angles is 180. Let x be that angle. Therefore, $x + 5x = 180$, $6x = 180$, divide both sides by 6: $x = 30$

47) Choice D is correct

Smallest 4–digit number is 1,000, and biggest 4–digit number is 9,999. The difference is: 8,999

www.EffortlessMath.com

... So Much More Online!

✓ FREE Math lessons

✓ More Math learning books!

✓ Mathematics Worksheets

✓ Online Math Tutors

Need a PDF version of this book?

Visit www.EffortlessMath.com

Visit www.EffortlessMath.com

for Online Math Practice

Made in the USA
Middletown, DE
18 September 2021